T0123019

EQUIPPING JAMES BOND

Equipping James Bond

Guns, Gadgets, and Technological Enthusiasm

André Millard

JOHNS HOPKINS UNIVERSITY PRESS BALTIMORE

Printed in the United States of America on acid-free paper
9 8 7 6 5 4 3 2 1

Johns Hopkins University Press
2715 North Charles Street
Baltimore, Maryland 21218-4363
www.press.jhu.edu

Library of Congress Cataloging-in-Publication Data

Names: Millard, A. J., author.
Title: Equipping James Bond : guns, gadgets, and technological enthusiasm / André Millard.
Description: Baltimore : Johns Hopkins University Press, 2018. | Includes bibliographical references and index.
Identifiers: LCCN 2018004446| ISBN 9781421426648 (hardcover : alk. paper) | ISBN 9781421426655 (electronic) | ISBN 1421426641 (hardcover : alk. paper) | ISBN 142142665X (electronic)
Subjects: LCSH: Espionage—Technological innovations. | Bond, James (Fictitious character)
Classification: LCC UB270 .M545 2018 | DDC 327.12028/4—dc23
LC record available at https://lccn.loc.gov/2018004446

A catalog record for this book is available from the British Library.

Special discounts are available for bulk purchases of this book. For more information, please contact Special Sales at 410-516-6936 or specialsales@press.jhu.edu.

Johns Hopkins University Press uses environmentally friendly book materials, including recycled text paper that is composed of at least 30 percent post-consumer waste, whenever possible.

CONTENTS

List of Abbreviations vii

Introduction 1

001 The Technological Enthusiasts 5

002 The Secret Intelligence Service 16

003 The Great War and the Threat of Modernity 27

004 Imagining the Future: Technology on Film 39

005 Spy Films 50

006 Ian Fleming, Intelligence Officer 63

007 Equipment 74

008 Irregular Warriors 86

009 The Treasure Hunt 95

010 Nuclear Anxieties 108

011 Gadgets 118

012 Guns 131

013 The Special Relationship and the Cold War 142

014 The Technological Revolution 153

015 Into the Future 165

016 Keeping Up with the Times 177

Notes 191

Index 207

ABBREVIATIONS

CR *Casino Royale* (New York: MJF Books, 1953)

DAF *Diamonds Are Forever* (1956), in *More Gilt-Edged Bonds* (New York: Macmillan, 1965)

DN *Doctor No* (New York: Macmillan, 1958)

FRL *From Russia with Love* (New York: Macmillan, 1957)

FVK *From a View to a Kill* (1960), in *Bonded Fleming* (New York: Viking, 1965)

FYE *For Your Eyes Only* (1960), in *Bonded Fleming* (New York: Viking, 1965)

GF *Goldfinger* (New York: Macmillan, 1959)

HR "The Hildebrand Rarity" (1960), in *Bonded Fleming* (New York: Viking, 1965)

LLD *Live and Let Die* (1954), in *More Gilt-Edged Bonds* (New York: Macmillan, 1965)

MGG *The Man with the Golden Gun* (New York: New American Library, 1965)

MR *Moonraker* (1955), in *More Gilt-Edged Bonds* (New York: Macmillan, 1965)

OCT *Octopussy* (New York: New American Library, 1966)

OHM *On Her Majesty's Secret Service* (1963), in *James Bond Omnibus*, vol. 2 (New York: MJF Books, 1992)

RSC "Risico" (1960), in *Bonded Fleming* (New York: Viking, 1965)

SLM *The Spy Who Loved Me* (New York: Viking, 1962)

TB *Thunderball* (1961), in *James Bond Omnibus*, vol. 2 (New York: MJF Books, 1992)

YLT *You Only Live Twice* (1964), in *James Bond Omnibus*, vol. 2 (New York: MJF Books, 1992)

EQUIPPING JAMES BOND

Introduction

James Bond surely needs no introduction. It is estimated that half the world's population has seen at least one James Bond film. Not since Sherlock Holmes has there been a fictional figure who has managed to free himself so completely from his creator to enjoy his own privileged position in popular culture. James Bond was a hero of espionage fiction well before he became a film commodity, but it is the 26 films (as of this writing) that have created the Bond we know today, as well as the Bond brand. In his study of the Bond films James Chapman says that they have become something of a ritual, and other commentators have argued that watching Bond films is comparable to a religion.[1]

Central to our enjoyment of the films are the gadgets, and it is significant that the actor who has achieved the longest run in the series is the man who equips Bond. Q has been called the Merlin to Bond's Arthur, and Desmond Llewelyn played him in 15 films from 1963 to 1999. The gadgets are so central to the Bond character that they can be used, along with what the press calls "the Bond girls," as a shorthand for each film in the series. In the opening titles of *On Her Majesty's Secret Service* (1969), which had the difficult job of introducing a new actor playing Bond, the gadgets are used to provide a history of the previous Bond adventures: the watch-garrote employed in *From Russia with Love* (1963), the underwater breathing device from *Thunderball* (1965), and the knife belt worn by Honey Ryder in *Dr. No* (1962). While Fleming's nostalgia for the disappearing past of secret intelligence survived in the character of James Bond, the nostalgia of the films' producers, Albert R. Broccoli and Herschel "Harry" Saltzman, for the halcyon days of "Bondmania" was expressed in the gadgets. The most famous of them all, the Aston Martin DB5, was retired as Bond's transportation after *Thunderball,* but it returned at the end of the century in two of Pierce Brosnan's Bond films. When Eon Productions (who owns the film rights and made all except two of the Bond films) decided to break with the past in the form of a new Bond—a more sensitive, modern man—the company wisely kept a link to Bond tradition in the form of the vintage DB5, which delighted viewers again in the 2006 remake of *Casino Royale.*

Ian Fleming insisted that his Bond stories were fantasies, yet he was one of the first novelists to include technical details about the equipment of espionage, which gave his stories some authenticity. Like the rest of his wartime peers, Fleming called this special equipment "toys," "gadgets," or "gimmicks," and he maintained that it was these rather than the sex and violence of the plots that his readers remembered. The gimmicks of the novels became the gadgets of the films, and they have become as important as the women and the martinis to the Bond brand.

So a James Bond without his little technological helpers would be unthinkable. In fact, such is the audiences' fascination with the gadgets that numerous books and articles have listed them and explained their use. There have also been several exhibitions of these artifacts, and both the Imperial War Museum and the Science Museum in London have used them to reflect on the development of technology in the twentieth century. This book takes the long view of the technological development of modern spyware and finds its roots in the Second Industrial Revolution of the late nineteenth century. Frightening advances in military technology pushed the British government into creating Bond's secret service, and from then onward the focus of espionage was closely related to the advance of military technology. James Bond was a product of World War II, and his equipment reflects that technological era and the tension between pure science and the ingenuity of amateur inventors.

Bond's equipment was drawn from the quartermasters of the Special Operations Executive (SOE) and other wartime commando and espionage organizations, as was his worldview and modus operandi. Yet between the publication of the first Bond novel, *Casino Royale,* in 1953 and the (second) film of the same name in 2006, the world and its technology changed drastically. As a modern and modernizing hero, James Bond has to keep up with the times, and the film franchise is committed to incorporating the latest technology into his gadgets and also into the armories of the villains he confronts. In this way the Bond films have become a showcase for stylish, modern technologies, taking the place of science fiction novels and World's Fairs of the twentieth century in anticipating the machines of the future.

Bond's global audience has often had difficulty negotiating between the real and the fantasy in his world. After the release of *Goldfinger* (1964) United Artists received letters complaining that an English film crew should not have been allowed into the inner sanctum of Fort Knox—which was, in fact, a creation of Eon's set designers. Although we in the audience realize that the plots are absurd and that the ending is predetermined, we tend to give the gadgets the benefit of the doubt because we have come to recognize Bond as a purveyor of advanced technology. The equipment of secret agents is an example of how art—in the form of motion pictures—has influenced life, and this book uncovers the prehistory of the Bond franchise by following the evolution of the spy or secret-agent film. The silent-picture fantasies of directors like Fritz Lang had a real impact on contemporary science and technology, and the equally fantastic equipment of James Bond has also influenced those involved in the business of espionage. The hidden knife blade in the evil Rosa Klebb's shoe in *From Russia with Love,* a gadget that attracted the attention of many in the intelligence community, did not have a World War II pedigree, nor was it a confidence shared with Ian Fleming by one of his friends in the British secret service; instead, it was the work of art director Syd Cain, who added a spike to the right shoe and built in a spring mechanism to make it pop out. Eon Productions has proudly claimed to have turned science fiction into science fact in the Bond franchise, so it seems premature to dismiss Bond's gadgets as fantasies when some of the wildly futuristic technology displayed in previous films, such as the ubiquitous smart phone, are now part of everyday life.

James Bond and his creator were both technological enthusiasts who embraced the new and who found beauty in machines, and Eon Productions has enthusiastically made the Bond films advertisements for photogenic new technology. At the same time, the villains and threats that Bond has to face have articulated contemporary fears about the downside of technological change. Born at the beginning of the Cold War, 007 now fights cyberterrorists and media moguls bent on world domination. The success of the Bond character is a result of maintaining a tradition and its rituals while burnishing every adventure with a layer of topicality and modernity. James Bond has survived by keeping up with the times, and each book and film is a product of its historical moment. Bond's wide appeal and the longevity of the brand make his equipment useful as a mirror of broader historical themes, especially public perceptions of the promise and threat of new machines. This book treats his gadgets as artifacts that, taken together, have acted as a barometer of hopes and fears concerning technology for more than half a century.

Perhaps the key to Bond's enduring appeal is that he maintains the ingenuity and resilience of an individual who faces a world brought to the brink of chaos by dangerous new technology. Bond will continue to take on these threats single-handedly and save the world until the final reel ends the cunning plans of the high-tech villains and sends the audience home comforted by this example of a lone hero triumphing over the machine.

Victorian ministers . . . let War pass out of the hands of experts and properly-trained persons . . . and reduced it to the disgusting matter of Men, Money and Machinery.

W. S. Churchill, *A Roving Commission*

001 The Technological Enthusiasts

Winston Churchill and Ian Fleming were perfect representatives of two generations of Englishmen—the Victorians and the Edwardians—who saw their world transformed. Churchill (born in 1874) and Fleming (born in 1908) lived through the Second Industrial Revolution, which brought a host of wonderful new inventions that changed life in ways large and small. The First Industrial Revolution (roughly 1760–1840) produced new power sources, new materials, and new products. The second came with the fruits of applied science and promised even greater social and economic transformations than the first. The philosopher Alfred North Whitehead argued that "the greatest invention of the nineteenth century was the invention of invention,"[1] and independent inventors like Thomas Edison and the research and development (R&D) laboratories of corporations like General Electric and DuPont produced a stream of innovations that persuaded the technological elite that they had the power to create a new world of productivity and affluence. Powered by electricity, integrated by mass communications, and colored by synthetic dyes, this technological revolution marked a high point in human endeavor, "an epoch of invention and progress unique in the history of the world," as one contemporary put it. Thomas Hughes called the century that began in 1870 "the era of technological enthusiasm," in which machine makers and system builders produced "goods for the good life," and technology transformed the material world. People had dreamed of utopias for centuries, but Howard Segal has described how new technology became linked with progress in the late nineteenth century. Technological enthusiasts believed that they had the cure for all of humanity's ills. For those who lived through it, the Second Industrial Revolution marked the beginning of a new modern age. The thinker and humanist Lewis Mumford saw it as the penultimate phase in technological development—the neotechnic era—in which the fruits of applied science had the potential to bring peace and prosperity.[2] For historians of technology there were new industries to study and new machines to admire. Reese Jenkins wrote about the emerging field of photography and how George Eastman brought it to the masses with his Brownie camera. This wonderful toy transformed our visual environment and made a permanent record of countless memories. Jenkins concludes: "One of the steadfast values of the last 200 years in America is that technical change is good."[3]

Churchill was born into a world with no cars or telephones. Flying through the skies in heavier-than-air machines or traveling underwater in submarines were dreams of science fiction. The Royal Navy, which was to play an important part in both Churchill's and Fleming's careers, was still sailing around the world in wooden ships in 1901, when HMS *Discovery* took Captain Scott's historic expedition to the Antarctic. By the time of Fleming's birth, humankind had taken to the skies, courtesy of the pioneering efforts of the Wright brothers and Von Zeppelin. The internal combustion engine and electric motor had revolutionized the application of power and the whole idea of mobility. In Ian Fleming's world telephones, electric lighting, and cameras were commonplace—essential parts of modern life, along with phonographs, typewriters, and motion pictures. Looking back on the "vanished age" of his youth, Churchill wrote: "The character of society, the foundation of politics, the methods of war . . . the scale of values, are all changed, and changed to an extent I would not have believed possible in so short a space."[4]

Although born thirty-five years apart, Churchill and Fleming led lives that were similar in many ways. Fleming admired Churchill and liked to think that both men had been considered black sheep at some stage in their careers. Ian Fleming might have been a child of the twentieth century, but he was still wedded to his country's nineteenth-century past and Churchill's Victorian value system. As Umberto Eco has pointed out, Fleming's "militaristic and nationalistic ideology, his racist colonialism, and his Victorian isolationism are all hereditary traits."[5] Born, like Churchill, into a good family (but perhaps not quite as grand), Ian Fleming was brought up in the same upper-class environment of privilege. Both men attended private schools and the Royal Military College at Sandhurst, and both indulged in some adventurous journalism as preparation for a life of writing. Churchill joined the Fourth Hussars and served as a young cavalry subaltern on the Indian frontier in the last years of Victoria's reign—surely the quintessential imperial experience: the officers' mess, the parades of gorgeously attired horsemen, and the games of polo, which Churchill and his comrades considered the "serious purpose of life."[6]

Both Churchill and Fleming were technological enthusiasts—gentlemen tinkerers who not only marveled at the complexity and beauty of new machines but were convinced that they had the power to change history. Although trained exclusively in the classics, they exhibited a practical, mechanical bent, using their eyes and their hands to understand the technology that was changing their professional lives, as well as their daily routines. In Churchill's early years a scientist was more likely to be an aristocrat or a well-established clergyman, an amateur investigator of the natural world, and the term *scientist* was not in general use. In Victorian England technological enthusiasm was generally confined to the classes who had the time to ponder the implications of the steady stream of scientific information or could afford the gadgets of the Second Industrial Revolution, but the introduction of the pneumatic tire changed all that.

The modern safety bicycle ushered in a new era of mobility for all classes and enticed a generation of young working-class men into a world of mechanical tinkering: repairing existing machines and designing new ones; developing the craft skills of turning, welding, and smoothing metal; and putting down their ideas in blueprints and drawings. They turned bedrooms into laboratories, built workshops in garden sheds, and founded small businesses in their homes. The hero of Wells's *The War in the Air* is Bert Smallways: "a vulgar little creature, the sort of pert, limited soul that the old civilisation of the early twentieth century produced by the millions in every country of the world . . . the sort of man who had made England and America what they were." Bert works in a bicycle-repair shop, and his life changes when he acquires a motorcycle: "So Bert grew up, filled with ideals of speed and enterprise, and became . . . a kind of bicycle engineer of the lets-'ave-a-look-at-it and enamel chipping variety."[7] The Bert Smallwayses of the world took the know-how of repairing bicycles into motorized transport, embracing the internal combustion engine to build motorbikes, boats, and cars. Some of them, like the Farman and Wright Brothers, applied their mechanical skills to powered flight, while others became advocates for the new mechanical age. Alfred Harmsworth was captivated by bicycling and leveraged his enthusiasm to become editor of the *Bicycling News.* He built a publishing empire of weekly magazines, popular newspapers like the *Daily Mail* and the *Daily Mirror,* as well as the venerable *Times* and *Observer.* As Baron Northcliffe he exercised considerable political influence and was one of the leading voices of technological enthusiasm in the United Kingdom.

A bicycle awakened Frederick W. Winterbotham's interest in technology, for this was the apex of mechanical travel at the turn of the century, but the bicycle was soon overtaken by more powerful machines, and Winterbotham graduated to a motorbike and then to a Studebaker automobile. Like many of his peers he was also fascinated with wireless communication. By 1901 Guglielmo Marconi was sending radio messages across the Atlantic, and boys like Winterbotham were building their own radio receivers and listening in to a new world in the ether. In Canada William Stephenson built his own Morse transmitter and tapped out messages to vessels on the Great Lakes. Stephenson was also drawn to aviation, making kites and model airplanes. His father had died in the Boer War serving with the Manitoba Transvaal Contingent, and during the next two world wars William served the British Empire, too. He became a good friend of Ian Fleming, who once said that while Bond was a fictional, romanticized spy, Stephenson was the real thing.

Many Edwardian technological enthusiasts were self-taught, and others benefited from the modernization of education, which created technical colleges where young men could take classes in electrical engineering or industrial chemistry. Geoffrey de Havilland (born 1882) received his engineering training at the Crystal Palace School, where he received a valuable patent for an improved motorcycle engine. He got a job at the Wolseley Company—one of the first automobile manufacturers. Frederick Handley-Page (born 1885) used his education at Crystal Palace to become chief designer at an electrical engineering firm, but after joining the Royal Aeronautical Society, he turned his attention to "aeroplanes."

Intoxicated by speed on land, water, and eventually in the air, these mechanically inclined young men were drawn to the wonders of the internal combustion engine. Like Toad of Toad Hall in Kenneth Graham's *The Wind in the Willows*, their moment of epiphany came when they encountered a motorcar. Winterbotham remembered the impact of these machines in England at the turn of the century: "It gave us the greatest excitement, when we saw the cloud of dust in the distance, to rush to the gates of our little house and watch the motorcars go by." Geoffrey de Havilland found this "new and exciting" technology irresistible and after his first ride in a motorcar stated, "I knew that my future life lay in the world of mechanical travel."[8] One of the pioneers of automobiling in the United Kingdom was Mansfield Cumming. Born in 1857, Cumming was brought up in a family of engineers, for both his father and brother were military engineers. He was an early member of the Automobile Club of Great Britain, and his passion for automobiles led him into the sport of motor racing. In 1903 he drove his 50 hp Wolseley car in the Paris-Madrid road race along with daredevil motorists like John Moore-Brabazon and Charles Rolls. The latter was one of the first in the country to own a motorcar and the first to open a car dealership. His meeting with the engineer Henry Royce in 1904 started a partnership that would define luxury automobiles in the twentieth century.

Winston Churchill was an early supporter of motoring, buying a Mors motorcar (produced by the French automotive pioneers Louis and Emile Mors) in 1900. He joined the Royal Automobile Club and soon bought a much larger Mercedes. As his income grew with his reputation as an author, he purchased several more expensive and faster cars, including a red Napier "Landaulette" for £580, a Rolls-Royce Cabriolet for £2,250, and a Wolseley sports car. Ian Fleming also sank his book advances and royalties into automobiles. He got his first taste of motor racing when he took part in the Alpine motor trials in 1932 as a reporter for the Reuters news agency. These trials ran on 1,500 miles of challenging roads across Germany, Italy, Switzerland, and France. When he returned to England, he bought a modest Standard Tourer, which he eventually crashed at a train crossing. He loved speed and recalled the thrill of reaching 100 mph in a three-liter Bugatti on a road near Henley. Fleming's cars got grander and faster as his fame as an author grew. After running a red Graham Page, he bought a 2.5 liter black Riley when *Casino Royale* came out in 1953. After securing the deal to sell the film rights to *Casino Royale,* he bought a Ford Thunderbird: "The engine, a huge, adapted low revving Mercury V-8 of five liter capacity, never gives the impression of stress or strain. . . . You can do a hundred [mph] without danger of going off the edge of this small island." As the Bond films brought him more money, he upgraded to a four-door T-Bird with a seven-liter engine and then a supercharged Studebaker Avanti—"a bomb of a motor"—designed by Raymond Loewy, who also produced the Studillac (a Studebaker with a Cadillac engine), whose speed and power James Bond praised in *Diamonds Are Forever.* Bond also dabbled in the world of racing and owned some powerful cars, and Fleming makes it clear that driving fast is a vital skill for 007: "These are Secret Service thrillers in which the hero and other characters make frequent use of fast cars and live in what might be described as 'the fast car life.'"[9]

His friend William Stephenson said that Fleming "was always fascinated by gadgets," and as a young man he was convinced of the powerful influence of technology on history.[10] During his preparation to take the Foreign Service exams in the 1920s, Fleming read up on subjects not normally taught at English schools: social history, anthropology, and the history of science and technology. When he had made a little money through banking deals in the early 1930s, he enlarged his book-buying hobby into a plan to build a library of technical and intellectual developments since 1800, "the milestones of human progress," as he put it. He instructed his book buyer Percy Muir to look for "books that have started something."[11] His collection started with Darwin's *On the Origin of Species* and Niels Bohr's *Quantum Theory* and then extended to books about motorcars, miners' lamps, zippers, and tuberculosis. Muir bought Fleming Madame Curie's doctoral thesis, about the properties of radium, and Sigmund Freud's book on the interpretation of dreams. Fleming poured money into his library and by 1939 could claim that the "Fleming Collection" was "one of the foremost collections of scientific and political thought in the world."[12] Collecting rare books was a hobby, but he also recognized that information about science was useful in his career goals in finance or journalism. He believed that technology made things happen, a sentiment shared by those of his generation who saw innovation as the driving force of social, economic, and political change. Fleming's conversion to technological determinism reflected the increasing usage of the term *technology,* which made it, in Eric Schatzberg's words, "a central keyword of late modernity," which was based on its association with applied science and the forces of modernity but encompassed both utopian expectations and dystopian fears.[13]

Schatzberg has described two distinct visions of technology: the instrumental view and the cultural view. The former sees it in terms of the autonomous power of the machine to affect history; the latter stresses the role of human values and agency in directing change. In *Technics and Civilization* Lewis Mumford rejected technology as the instrument for "material conquest, wealth, and power" and hoped that "the machine . . . will fall back into its proper place: our servant, not our tyrant."[14] Fleming's admiration for fast cars, guns, and gadgets put him firmly in the former definition of *technology,* and his fiction made these boys' toys into fetish objects that had the power to shape the narrative. Fleming's belief in the efficacy of the machine and his understanding of technology as the driving force of both modern history and his plots did not ignore the role of human agency, but instead of enlightened government regulation, responsible scientists, and socially conscious engineers, he saw redemption in the form of one well-equipped secret agent.

In 1908, the year of Fleming's birth, the first flight was undertaken in the United Kingdom by the American showman Samuel F. Cody in a craft modeled on the Wright Flyer but powered by a 50 hp Antoinette engine. At this time airplanes were a marvelous curiosity. Winterbotham recalled: "It must, I know, be surely difficult in these days of universal air travel to realize what flying meant to a young man of nineteen some seventy years ago. It was [as] if a new dimension had suddenly been added to life."[15] Over in France Wilbur Wright's demonstrations of the Flyer at Le Mans created a sensation. He wrote home to his brother: "the newspapers continue exceedingly friendly and the public interest and enthusiasm continues to increase."[16] Subsequent flights over Europe drew crowds of thousands, including Lord Northcliffe and King Edward VII, and inspired a generation of technological enthusiasts to take to the air. One of them was Geoffrey de Havilland, who merely read about Wright's demonstration but "though I might never have seen an aircraft in the air, this was the machine to which I was prepared to give my life."[17] He designed his own airplane and sold it to the government balloon factory at Farnborough, which later became the Royal Aircraft Factory. Handley Page lost his job at Johnson and Philips electrical engineers because he spent too much time in aviation experiments, and he left to form his own concern to build airplanes. John Moore-Brabazon (born 1884) bought a French Voisin-Farman airplane and took it up in 1909. For this feat he was awarded the first pilot's license issued in Britain, and he was soon followed by other gentleman aviators such as Thomas Sopwith. An ardent motor racer and motorcyclist, Sopwith purchased a craft built by Howard Wright, taught himself to fly, and achieved a record-breaking flight of 107 miles in 1910. He then opened a flying school at Brooklands and began to build his own aircraft. Moore-Brabazon joined with other enthusiasts like Mansfield Cumming and Charles Rolls, who were both founding members of the Royal Aero Club. Cumming gained his pilot's license in 1913 in a Farman biplane at the ripe old age of 54, and Rolls bought a Wright Flyer made under license in England and flew the first east-bound crossing of the English Channel with it.

Flying appealed to Churchill, the technological enthusiast: "From the outset I was deeply interested in the air and vividly conscious of the changes it must bring to every form of war." As early as 1909 he was pressing government officials to approach the Wright brothers to jointly develop the military application of their invention, and he was a vocal proponent of creating a "Corps of Airmen to make aviation for war purposes." He took his first flight around 1911 and was captivated by "excitement and curiosity. . . . I continued for sheer joy and pleasure. I went up in every kind of machine and at every air station under the Admiralty."[18] After being appointed First Lord of the Admiralty in 1911, Churchill embarked on a program of modernization, which included the significant shift from coal to oil-fired ships and the establishment of the Royal Naval Air Service (RNAS). Churchill was a frequent visitor to the RNAS depot at Eastchurch and flew in a variety of seaplanes and airships. He also took advantage of the training programs he set up to learn how to fly under the instruction of Captain Ivon Courtney, RNAS, and Captain Alan Scott at the Central Flying School. Churchill's flying lessons were curtailed because of the obvious dangers of flying (Captain Scott reported that Churchill was slightly injured in an accident) and the entreaties of his wife and colleagues. He never obtained his certificate, yet his interest in aviation continued.

By 1908 both automobiling and aviation were well established, and several entrepreneurs had founded companies to make these wonders available to the adventurous. In 1908 Henry Ford introduced his Model T, and Messrs. Rolls and Royce built their factory in Derby to manufacture their "40/50 hp" model, which, as the Silver Ghost, was called "the best car in the world" by *Autocar* magazine. In the same year, Orville Wright completed flights of more than an hour and took the first passenger aloft. In 1908 it was possible to purchase an aircraft from the Wright brothers, from the Voisin aviation company, and from Maurice and Henry Farman. The Wright brothers had entered into a contract with the US Board of Ordnance and Fortification to supply "an operative machine capable of carrying two persons" and were marketing their improved Flyers all over Europe. At a demonstration at Le Mans, France, the Flyer caught the attention of a man who called himself Sidney Reilly. Reilly made an agreement with the Wright brothers to be their sales agent in the Russian Empire. While Wilbur Wright traveled across Europe publicizing heavier-than-air craft, Count Ferdinand von Zeppelin was touring the skies in a rigid airship filled with gas bags. In 1908 his fourth experimental craft managed to stay aloft for 24 hours, and he claimed that airships would be the future of commercial and military aviation. Newspaper tycoons and technical societies were offering prizes for record-breaking flights. *Scientific American* presented a silver statuette for the first American flight of one kilometer, and in 1908 Lord Northcliffe offered a prize of ten thousand pounds to the first man to fly across the English Channel. Louis Bleriot flew across the Channel the next year, winning the prize and earning his place in history.

Not everyone in Great Britain welcomed Bleriot's flight with enthusiasm. The *Daily Express* announced that "Britain was no longer an island," and other newspapers castigated British science and industry for failing to keep up with the French. H. G. Wells wrote a letter to the newspapers in which he blamed the British education system for lagging behind in this significant new technology, which might seriously threaten the security of the kingdom. While the accomplishments of two industrial revolutions convinced many educated men of the unlimited potential of human ingenuity, there were those, like H. G. Wells, who saw the dangers of letting the genie out of the bottle. He called the Second Industrial Revolution "The Scientific Age" because in his view science "altered the scale of human affairs" and eroded the old national boundaries and social conventions, uprooting men from the soil and exposing them to a "torrent they never clearly understood." This was the same "technological torrent" that historian Perry Miller evoked as he described how Americans flung themselves into it, "shouting with glee . . . as they went headlong down the chute that here was their destiny." John Ellis has emphasized "the absolute faith that men in the nineteenth century had in the beneficial effects of scientific, technological and industrial progress," when even the inventors of machine guns claimed that these deadly weapons "are intended to strike terror into the hearts of every enemy . . . which will create an enthusiasm and a sense of security in every nation on this globe."[19]

The technological enthusiasts accepted that their wonderful machines were also dangerous. One of the adventurous young men Cumming employed reported that he had "given me some hairy rides," and in 1914 Cumming collided with a wall at about 70 mph (an amazingly high speed for the time), killing his son and costing him part of his leg. Aviation was even more dangerous. The first casualty was Captain Thomas Selfridge, a passenger in a Wright Flyer that crashed at Fort Myer in 1908. Charles Rolls died in an air crash in 1910, another first for an Englishman, and others walked away from bad accidents. Geoffrey de Havilland survived one that destroyed his biplane. By 1910, 26 daring young men had died in air crashes. As the leading aviation enthusiast in the United Kingdom, Moore-Brabazon accepted the dangers of flying: "I think we were all a little mad, we were all suffering from dreams of such a wonderful future."[20]

The concerns about the dangers of new machines were first focused on military technology. The first aircraft to go aloft in the United Kingdom was flown by an American showman but had been co-designed by a Royal Engineer and designated "British Army Aeroplane Number 1." The disruptive potential of new military technology had become evident in the late nineteenth century when mass armies equipped with breech-loading rifles, rapid-fire artillery, and machine guns inflicted unprecedented destruction on the enemy. The new artillery pieces could fire much farther, so the need arose for long-distance observation, which was met by balloons connected to the ground by electric telegraphs. In the Franco Prussian War of 1870–71 the German army used its train network and advanced weaponry to route the French with a speed that surprised other global powers. The standard design of the handgun—a revolver with a cylinder holding five or six bullets popularized by Samuel Colt—was replaced by an automatic pistol that employed the recoil or gas blowback to eject the spent cartridge and insert a new one—the same idea behind the machine guns invented by John Browning and Hiram Maxim. Browning's design for an automatic pistol was taken up by Fabrique Nationale of Belgium, and it was one of their guns, an FN 1900, that was used by Gavrilo Princip to assassinate

Archduke Franz Ferdinand in Sarajevo in June of 1914—the first shot of World War I.

For technological determinists there was no better justification for their beliefs than the battlefield. Churchill fought in the Battle of Omdurman in the Sudan in 1898, and his heroic account of the battle was widely read. Yet his description, centered on the charge of the 21st Lancers in which he took part, omitted the role played by six Maxim machine guns. Another observer noted the effect they had on the advancing Dervish forces: "It was not a battle but an execution." With fewer than 50 casualties, the British claimed to have killed about 10,000 tribesmen. In his autobiography Maxim proudly reproduced this conclusion by Sir Edward Arnold: "In most of our wars it has been the dash, the skill, and the bravery of our officers and men that have won the day, but in this case the battle was won by a quiet scientific gentleman living in Kent."[21]

In 1906 the Royal Navy introduced a new battleship called the Dreadnought, an all heavy-gun ship powered by steam turbines and equipped with the largest guns yet put to sea. It was aptly named, for it was so much faster and more powerful than any other ship afloat that it made all existing battleships obsolete overnight. It started an arms race of the industrial powers to build bigger, faster, and more powerful ships—the Super Dreadnoughts—"each larger and heavier and more deadly than its predecessors. Each in its turn was hailed as the last birth of time," wrote H. G. Wells, who decried the outrageous cost of these massive ships. The companies building these vessels connived with a pliant press and politicians to make the Dreadnought a symbol of a nation's technological prowess and military power. Images of these vessels were put on postcards, cigarette packages, crockery, toys, and prints to be framed and exhibited in living rooms. For technological enthusiasts the great ships were "tremendous engines of war" (Churchill); for its detractors they were "strange monsters" (Wells). Fred Winterbotham was mesmerized by a review of the fleet: led by "the great battleships like Dreadnought with their vast 16 inch guns . . . the spectacle on that hot summer's day was one of the most memorable of my young life."[22]

Yet no wonder weapon was safe from the challenge of new machines, which threatened an even greater disruption of the balance of military power. The first submarines had been introduced during the American Civil War, but pioneer boats like the CSS *Hunley* were more dangerous to their crews than to surface ships. John Phillip Holland was a schoolteacher and amateur inventor who approached the US Navy in 1875 with plans for a submersible fighting ship. Rejected by the navy, Holland took his plans to Irish republicans as a counterbalance to the battleships of the Royal Navy. The Fenians funded Holland's research and his first submarine, the 14-foot-long *Fenian Ram,* was launched in 1881. This was the first modern submarine. The Holland VI was finally accepted by the US Navy in 1900. It had a dual propulsion system (diesel engines for surface cruising and electric motors for underwater) and could fire a single torpedo. Holland's improved boats were made under license and purchased by many naval powers.

Spectacular innovation in military technology inevitably led to arms races. After completing the first generation of Dreadnoughts, the German navy's construction department recommended a new series of bigger and more powerful ships in 1908. Acquiring the patents for the Parson's turbine meant that the Kaiser, Konig, and Bayern classes would match any British warship. Germany's navy also took an early interest in von Zeppelin's rigid airships, and by 1914 it had a fleet that the Royal Navy correctly considered an offensive weapon aimed at the United Kingdom. A few years earlier H. G. Wells's readers had been confronted with a story that anticipated the use of airpower to destroy great cities and a world war initiated by an aggressive and imperialistic Germany. *The War in the Air* (1908) described an attack by a fleet of German airships on the United States that first destroys the Americans' fleet of Dreadnoughts at sea (proving that sea power is no match for airpower) and then proceeds to bombard New York. In his description of aerial attacks and a mechanized war that destroys civilization, Wells started on a theme that he continued in *The Shape of Things to Come* and anticipated much of the destruction of World War II.

Churchill and Fleming also recognized that a future of technological advancement could overturn existing values and hierarchies. As a student undergoing the training course at the Royal Military College, Cadet Fleming resigned in 1927 and later said: "I didn't become a soldier after passing out from Sandhurst because they suddenly decided to mechanize the Army and a lot of my friends and I decided that we did not want to be glorified garage hands—no more polo, no more pig sticking and all that jazz." This sentiment was echoed by Churchill. Looking back from the 1930s, he concluded: "It is a shame that war should have flung all this aside in its greedy, base, opportunist march and should turn instead to chemists in spectacles, and chauffeurs pulling the levers of aeroplanes or machine guns. . . . War, which used to be cruel and magnificent, has now become cruel and squalid. In fact it has been completely spoiled. It is all the fault of Democracy and Science."[23]

The feeling of danger from fresh discoveries affected the patriotic imagination of every people in the world.

H. G. Wells, *The War in the Air*

002 The Secret Intelligence Service

The Secret Intelligence Service (SIS) that Ian Fleming made famous was created the year after his birth. In August 1909 naval commander Mansfield Cumming was asked to visit Rear Admiral A. W. Bethel, the Director of Naval Intelligence (DNI), a post that had been created in 1887 along with the Director of Military Intelligence. Although both of the armed services had intelligence departments, the navy had by necessity a more global outlook and was facing the disconcerting pace of German naval rearmament. Cumming understood that the "duties of SS Bureau" were to "organise an efficient system by which German progress in Armaments and Naval construction can be watched."[1] One of the earliest operatives in this new department, Hector Bywater, identified by his designation H2O, "sent a lot of good stuff" on German shipbuilding and submarines. His report on big naval guns brought him to London to be interviewed by the admiral in charge of naval gunnery. Aeronautical matters were another priority, and H2O frequently reported on the construction of airships in Germany. He also recalled that the fleet paymaster Charles Rotter "asked me to get him information about secret building of Submarines."[2] H2O's information often failed to convince a naval staff that harbored doubts about the usefulness of such weapons in the first place and was patriotically unwilling to concede that the Germans had achieved technological leadership in the submarines, torpedoes, and dirigibles that were going to transform war at sea.

The technological advances of the Second Industrial Revolution transformed the objectives of the "Great Game" of international intrigue described by authors like Rudyard Kipling and John Buchan. Intelligence gathering had not changed much during the nineteenth century, when spies seduced foreign diplomats or set out in disguise to sketch enemy forts or investigate dockyards. Robert Baden-Powell published his exploits as a spy, which read like adventure stories for boys: hoodwinking a Turkish officer to let him examine a fort and then "I jotted down on my shirt cuff . . . the information required by my superiors at home," camouflaging plans of enemy installations as drawings of butterflies, and sections on "Secret Signals and Warnings" and "How to Enter a Fort."[3] But the new weapons systems generated by the arms races focused espionage onto military technology. In Sherlock Holmes's day the prize was often diplomatic documents, such as the Naval Treaty, "a document of immense value," as Watson put it in an adventure that William Baring-Gould dates to 1889. *The Adventure of the Second Stain* (1886) also dealt with an incriminating document, this one indiscreetly written by a foreign potentate. Yet the papers that the French army officer Alfred Dreyfus was accused of stealing and selling to the Germans in 1894 were not about the French army's order of battle, nor their plans to invade Germany, but the development of a new quick-firing artillery piece. The attention of intelligence organizations was being drawn to the laboratories and drawing offices of the great armaments manufacturers like Krupp in Germany, Schneider-Creusot in France, and Vickers-Armstrong in the United Kingdom. Sidney Reilly's well-publicized, and often exaggerated, exploits as a "master spy" include his daring theft of secret plans from the laboratories of these concerns. By 1895 Holmes and Watson were involved in *The Adventure of the Bruce-Partington Plans,* which were the blueprints of a new submarine and the most "jealously guarded of all government secrets. . . . Naval warfare becomes impossible within the radius of a Bruce-Partington operation."[4]

The arms race with Germany created a new literary form: the paranoid spy novel that married the obligatory tales of derring-do with the current anxiety about German expansionism and the secret weapons they might have. This type of story first appeared in the aftermath of the Franco-Prussian War, which had provided evidence not only of Germany's advanced weaponry but also of its highly efficient espionage service. Upright Britons like Baden-Powell imagined thousands of German spies at work, and its recent victory was bolstered by "over 20,000 paid informers stationed in France and controlled by one man."[5] In 1871 George Tomkyns Chesney's *The Battle of Dorking* described a surprise assault on England by German-speaking invaders. One reviewer wrote that "it describes exactly how we feel," and the book was such a popular success that hundreds of similar stories followed.[6] The book that has the best claim to be the first modern spy novel also articulated the threat of a German invasion of the United Kingdom. *The Riddle of the Sands,* by Erskine Childers, was published in 1903. It describes how Carruthers of the Foreign Office takes a boat trip along the Ems and Weser estuaries on the German coast and finds evidence of a planned

German invasion of eastern England. It was an entirely modern thriller with convincing technical detail about sailing and railways, and it impressed many who read it, including John Buchan, who thought it the best adventure story of its time. The leading exponent of the paranoid spy novel was William Le Queux, who had the support of Northcliffe and his newspapers. A series of Le Queux's stories that ran in the *Daily Mail* in 1906 were turned into *The Invasion of 1910,* which described a successful invasion of England. It quickly sold a million copies. In 1909 Le Queux published *Spies of the Kaiser,* which depicted them in large numbers all over the United Kingdom as they gathered information about the movements of the Grand Fleet and used wireless messages to send it back to Germany. The intelligence officer and civil servant John Buchan was serving in the War Propaganda Bureau when he wrote *The 39 Steps* in 1915. This is the extraordinary story of an ordinary civilian called Richard Hannay, who stumbles upon a ring of German spies and their plot to take vital information about the British Grand fleet out of the country. After a thrilling chase across the Scottish Highlands, Hannay thwarts them and saves the fleet.

In the years immediately before the Great War the British press was full of alarmist stories about threats of invasion and the presence of hundreds of German spies in England. The *Daily Mail* stoked the fires of anti-German sentiment, and the *Weekly News* even ran competitions to spot spies. The threat of German spies was taken seriously by the government. In 1911 Home Secretary Winston Churchill was surprised to learn that two navy arsenals were guarded by only a few constables and asked what would happen if "twenty determined Germans in two or three motor cars arrived well armed upon the scene one night." While serving as home secretary, Churchill was a vocal supporter of enlarging the British intelligence operation and instructed that the mail of suspected German spies was to be opened. As First Lord of the Admiralty in 1914, Churchill was part of a group of officers motoring in northern Scotland to reach the great naval base at Scapa Flow and the Dreadnoughts of the Grand Fleet moored there: "the whole of the War ultimately hinged upon this silent, sedulously guarded, and rarely visible pivot . . . on which the command of the seas" depended. In an account that could easily have been written by Buchan, Churchill noticed a large searchlight on a nearby castle and imagined a flotilla of German submarines sitting offshore, waiting to ambush the fleet, and a Zeppelin loitering close to the shore that would pick up the signal from the searchlight and relay the information by wireless to the submarines. His suspicions might have been inspired by spy novels but were informed by his insider knowledge of the capabilities of new military technology: the airship and submarine facilitated a new level of surveillance, at once more timely and more far-reaching. Churchill and his colleagues therefore approached this Scotch shooting-lodge armed with pistols, expecting to confront a nest of German spies.[7]

Fears about foreign espionage agents in England stirred up by Queux's vivid imagination had some influence on the decision to create a secret intelligence service. In 1908 James Edmonds of Military Operations sent a memo to the chief of the General Staff listing his concerns that the government was unprepared to counter the threat from German espionage. He pointed out that there was no permanent staff to watch suspicious cases and no cooperation among the departments charged with law enforcement, intelligence, and counterespionage. Later investigations carried by British intelligence indicated that German naval intelligence under the direction of Gustav Steinhauer had set up its espionage operation in the United Kingdom as early as 1905. The target was naval dockyards, weapons facilities, and codebooks. Around 1908 Steinhauer started a vigorous campaign to infiltrate Great Britain with spies, bringing them in from the Netherlands, Belgium, Norway, Denmark, and the United States. The convicted German agent William Klare told a fellow prisoner in 1912 that German spies were present in every rank of life in the United Kingdom. In response to the concerns raised by military intelligence, the Committee of Imperial Defence recommended that a "Special Intelligence Bureau" should be established "with the object of counteracting the efforts of the German government to establish a spy organization in the United Kingdom." This was carried out in October 1909. The new organization was to have two parts: home and foreign. The Home Section MO5 was put under Captain Vernon Kell (and was later renamed MI5), and the Foreign Section (which we now know as SIS or MI6) was given to Mansfield Cumming. Edmonds's fears were soon justified, and from 1911 to 1917 MO5 initiated proceedings against a steady stream of German agents, many of whom were caught seeking out secrets embedded in new naval and aviation weaponry. Spies were sent to find out about torpedo directors, fire directors, a new mercury sight designed to compensate for the rolling and pitching of a warship, mines, minesweepers, and wireless equipment. The Annual Report of Torpedo School was of special interest to German intelligence, and at least three agents were charged with acquiring a copy. Special attention was paid to naval aviation. Agents were sent to inspect "hydroplane" stations and find out about new aircraft being developed by military contractors like the Short aircraft company. The German intelligence service set up a clever scheme to pose as a publisher of an encyclopedia of naval technology and then press contributors (who were to be handsomely paid) to find out more about "novelties and modern inventions."

The British authorities even had to ban the printing and sale of the postcards that depicted Dreadnoughts and aeroplanes in case German agents sent them to friends and relatives abroad.[8]

The Secret Intelligence Service

The technology of the Second Industrial Revolution not only redirected the focus of intelligence gathering but also changed the way it was carried out. The tools of espionage had always been the sketch pad, a bottle of invisible ink, a decent telescope, and a suitcase full of disguises. Counterintelligence did not go much beyond shadowing shady characters and opening their mail. When Cumming sent one of his agents to find out if U-boats were being refueled off the Spanish coast, he gave the agent his "telephoto camera in two cases . . . a battery of lenses . . . two pairs [of] Zeiss glasses [binoculars] . . . [and] a Ross telescope." Another agent was equipped with a codebook, a safe, and a pair of Zeiss glasses. Cumming's agents sent back intelligence in the form of hand-drawn sketches and maps, converting them into tiny pieces of paper hidden in signet rings or shoes, or left them in hidden places—dead letter boxes—for other agents to pick up. They sent reports written in secret ink or smuggled them through enemy lines hidden in their clothing. Most of these secret messages were in code, some not, and some were in blank verse. The author Compton Mackenzie's literary efforts amused his boss, but the army intelligence officer who was the customer for this information requested "a simpler and less melodramatic literary style."[9]

Some of the gadgets produced during the Second Industrial Revolution—such as the phonograph, camera, and wireless set—could be applied to the work of espionage. The spy no longer had to depend on secret inks and the mail service to report back to base, for now he or she had photography, mimeography, radio, and telephony. The old-fashioned spy who had been identified by his or her handwriting on secret reports could now use a typewriter. Copying rather than stealing documents decreased the risk of discovery, and now there were several new technologies available to make copies. Electronic communication diminished the need for the risky personal interaction of agent and controller—but it also provided opportunities for the opposition to listen in. Technology was to change espionage in World War I as much as it was to transform the battlefield.

Communication was the key to modern warfare; orders scribbled by hand and sent by courier to commanders in the field were not fast or reliable enough to manage a modern battle, nor were flags or semaphore systems. One of the first offensive actions of the Great War was carried out by the Royal Navy when it cut Germany's telegraph cables, severing its communication with the rest of the world. This forced the Germans to use the cable network of neutral nations or send messages by wireless—both of which were vulnerable to interception. Each side set up listening stations and recruited clever men from their universities to break into the enciphered wireless messages that had been captured from the ether. British naval intelligence led the way into this new technology of espionage. The eminent scientist Sir Alfred Ewing was given command of a small department tasked with deciphering the growing number of intercepted messages. He recruited linguists, mathematicians, radio enthusiasts, and engineers from the communications industry to join him in Room 40 of the Admiralty. Under the leadership of DNI Reginald "Blinker" Hall, Room 40 achieved some of the greatest intelligence coups of the war. The decipherment of the infamous Zimmerman Telegram, which helped bring the United States into the conflict in 1917, was called "the most important enemy signal to be decoded in either World War."[10] This triumph established signals intelligence (SIGINT) as an important part of intelligence gathering.

Breaking the enemy's code required mathematical analysis, familiarity with the language of the sender, and big slices of luck. Room 40 benefited from a windfall of captured German codebooks in the last months of 1914. The German cruiser *Madgeburg* ran aground in the Gulf of Finland, and the Russians found a copy of the *Signalbuch der Kaiserlich Marine* in its chart room along with the current cipher key and many helpful maps. HMS *Theseus* was sent to collect this codebook, and it was personally handed over to First Sea Lord Churchill by Captain Smirnoff of the Russian navy. A few weeks later, a British trawler brought up a lead-lined chest containing a codebook that had been thrown overboard by a German destroyer during an engagement in the North Sea.[11] These finds proved to be so useful that Hall issued orders that enemy vessels sunk or captured should be searched for codebooks; then, he went one further by initiating special operations to acquire codebooks from German vessels by deliberately leading them into traps and sinking them. Navy divers on the scene then went down to retrieve the codebooks. Blinker Hall's men found good pickings in sunken German U-boats—an important precedent in the history of code breaking that did not escape the attention of Lieutenant Fleming of the RNVR in the next world war.

Aerial Reconnaissance

As war drew closer, European intelligence requirements shifted from finding out about the enemy's advanced weapons to determining exactly when and where he would attack. The writers of spy novels would have preferred that this task be undertaken by exotic female spies in the mold of Mata Hari, but there were more efficient and less glamorous ways to do it. As far as the army was concerned, the best intelligence it ever got from Cumming's department came from Belgian civilians, untrained in the dark arts of espionage, who simply counted German troop trains on the way to the front, providing an assessment of troop dispositions, reinforcements, and the direction of future attacks. On 13 August 1914 the Royal Flying Corps (RFC) sent a squadron of aircraft to support the British Expeditionary Force (BEF), and its first aerial reconnaissance was carried out a few days later by a Bleriot monoplane. These efforts soon proved their worth. When the BEF stopped the German invaders at the Battle of Mons, along the Mons-Conde canal, aerial observation spotted a flanking movement and provided Sir John French with the vital information to achieve a hasty withdrawal and thus save his troops. The intelligence provided by aircraft was accurate, wide-reaching, and, most important, timely. In the war of movement that marked the first months of the war, only aerial observation could provide up-to-date information on the rapidly changing battlefield. The prewar opposition to the use of aircraft from the commanders of the British army quickly evaporated, and the RFC presence in northern France was quickly built up.

Although some pilots grappled with the new skills required in aerial reconnaissance, others, like William Stephenson, quickly appreciated its benefits and used their craft to observe the enemy at every opportunity. No. 1 Wing of the RFC was created to provide reconnaissance to the BEF, and in January of 1915 a photographic unit was set up. The first observation flights depended on sketches and marks on maps, but it became clear very quickly that a camera was the best tool for the job. John Moore-Brabazon returned to flying, joining the RFC as a lieutenant, and his knowledge of photography brought him into the photographic unit, where he was joined by Lieutenant Charles Cambell and Sergeant-Major Victor Laws. These three were responsible for building up the aerial reconnaissance and photo interpretation operations. Brabazon and his colleagues designed a special camera for use in aircraft. The A-type was a handheld, wooden camera with leather bellows, which used 4×5 inch glass-plate negatives and was operated by the pilot or an observer who had to lean out of the cockpit to look through the ring-and-ball sight and pull a toggle to take the picture. Each picture required a new plate, which had to be loaded into the camera during flight—no easy matter. The improved C-type was attached to the airplane and could automatically load 18 glass plates from its magazine, providing a series of photographs that could be overlapped to produce three-dimensional images, which were especially important in targeting specific installations. True stereo cameras were also taken into the air, primarily by the Royal Naval Air Service (RNAS), which maintained a leadership in aerial technology. The E-type camera was made of metal and could be operated by remote control from the cockpit. The RFC used different lenses for low-level oblique and high-level shots. The A- and C-types had the standard eight-inch lens, but other cameras, such as the L-type, had much longer (up to 20 inches) lenses, enabling photography from higher and safer altitudes. By the end of the war the RFC had cameras that could produce acceptable images (which were clear enough to transfer into 1:20,000 scale maps) from 20,000 feet.

Hand in hand with the development of the aerial camera came parallel developments in organization and training: technological innovation depends on much more than new machines, which must be integrated into social organizations. Once the photographs were taken and developed, they had to be examined for information and the images interpreted. The School of Photography, Mapping, and Reconnaissance was set up at Farnborough in 1916, and many individuals contributed innovations that formalized photographic interpretation and set up procedures to develop maps from photographs. Pioneers like the barrister and surveyor Captain Charles Romer laid down the first rules for scaling air photographs.[12] The first manual, *Notes on the Interpretation of Aeroplane Photographs,* was introduced in 1916. By the end of that year, every squadron at the front was equipped with an intelligence section comprising an intelligence officer, two draftsmen, and a clerk. This section briefed and debriefed the crews and made the initial interpretation of the prints. Under the direction of Victor Laws, photo reconnaissance became a vital part of the Allied intelligence effort in the West.

As demands for photographs became insatiable, aerial combat over the front lines was concentrated on protecting or destroying observation aircraft. Winterbotham's unit had to accompany the camera planes, beating off the German scouts and attacking the enemy's reconnaissance planes. Winterbotham said that the army (which ordered the reconnaissance flights) "seemed blithely unmindful of the tremendous losses of highly trained pilots and valuable aircraft which air photography entailed, and which the Germans were anxious we should not carry out." Yet the photographs had to be taken at all costs: "Urgent demands came for photos, and more photos, to see what his [the Germans'] next stand would be." During the Battle of the Somme (July–November 1916) the RFC took more than 19,000 photographs, from which 430,000 prints were made. By 1918 both sides were recording the entire front every day, and by the time of the armistice in November the RFC had taken nearly six million aerial photographs that year.[13]

Technical Innovation in the Secret Service

The spy mania of the prewar years increased dramatically after August 1914. The eminent spy catcher Basil Thomson of Scotland Yard noted that "the malady assumed a virulent epidemic form accompanied by delusions which defied treatment."[14] Cumming was able to increase the size of his service, building up its headquarters to around 60 office workers and technical staff and buying special equipment. His associates remembered him as imaginative and open to innovation; slowed down by his wooden leg, he bought an "Autoped" scooter from the United States and used it to speed around the corridors of his office. One of his agents wrote: "He had a passion for inventions of all sorts, and being a rich man, he often bought the rights to them, such as strange telescopes, mysterious mechanisms with which to signal in the dark across the front, rockets, bombs etc." He set up a small workshop in his headquarters with a lathe "for bomb making etc." and brought his own tools to equip it.[15] SIS had a laboratory for experimenting with invisible inks and training agents in their use. After he was recruited, Paul Dukes was taken "to the laboratory to learn the inks and all that." Dukes described Cumming's office: "A row of half a dozen extending telephones stood at the end of a big desk littered with papers. On a side table were numerous maps and drawings, with models of aeroplanes, submarines and mechanized devices, while a row of bottles of various colors and a distilling outfit with a rack of test tubes bore witness to chemical experiments and operations."[16]

At the beginning of the war, the time-honored secret ink was lemon juice, and if a suspect was found with a small bottle of it along with a ballpoint pen or a toothpick, this was enough for MO5 to make an arrest. But the Germans had the benefit of advanced chemistry and produced improved inks, which could be concealed as scent or other liquids. Some ink was carried in the form of crystals hidden in cosmetics or tiny metal tubes. German agents carried small pieces of blotting paper or fabrics that were immersed in water to produce an ink that could only be revealed by a special developing process. They were told that the transmission of their intelligence was "the great danger" but that German chemical science had reduced that risk to a minimum by inventing inks that could not be detected.[17] Consequently, British censors and counterintelligence officers had to improve their methods of detecting secret inks. Once it was discovered that iodine vapor could develop secret writing, the censor's office built special containers to quickly apply this process to letters. Throughout the war each side pursued technological innovation to gain an advantage or to counter the enemy's advantage—a scientific contest that was to become a vital element of modern warfare.

Cumming, the technological enthusiast, sought out scientific expertise and was always on the lookout for special technical equipment. He hired the physicist Thomas Merton in 1916, the first scientist in the secret service who "worked at secret inks, bombs, etc." Cumming regularly consulted faculty at the University of London for information on secret inks. Keith Jeffery found no scientific or technical staff in the account books of SIS in the early 1920s, but there was a record of Cumming recruiting a chemist in 1919.[18] Cumming was an innovator in many fields, including recruiting women for intelligence work and in formalizing training, which had hitherto been learned on the job. He had a training pamphlet printed entitled *Notes on the Instruction and Recruiting of Agents.*

The secret service was moving with the times, rationalizing its organization and adopting new technologies like photography. A photograph was far more exact than a sketch; thus, when traitorous gunnery officer George Parrot obtained some navy codebooks for German intelligence, they were whisked over the Channel to Ostend and photographed by agent Robert Tornow (known as T), accompanied by two assistants and special appliances. Yet photography was also useful in counterintelligence. Vernon Kell ordered that photographs be taken of all suspected agents and then circulated among dockyards and other government agencies as a matter of routine. Along with a specimen of handwriting, photographs were part of every file of suspicious persons. Photographic evidence was also used in prosecuting them.[19]

Sound recordings played the same role as photography in secret intelligence in that a recording was an exact and unimpeachable transcription. The Dictaphone, used for recording dictation in offices, was an offshoot of Edison's early cylinder phonographs, but in intelligence work it recorded interviews and intercepted messages in addition to easing the burden of administering the intelligence organizations. As the size and scope of these organizations grew, so did the amount of data they produced. In 1914 MO5 had 17,500 card indexes; by 1918 it had more than 250,000.[20] A modern intelligence service is only as good as its methods of saving and accessing information, and by the end of the war the Secret Intelligence Service had taken steps to formalize its rules for saving and organizing data. The technology of espionage moved hand in hand with the development of office machinery because the telephone and copying machine were critical in gathering and distributing information. Office machinery also played an increasingly important part in running large organizations, some of which had spying as their principal business. In a major reorganization of MO5 in 1917, three new sections were established within G branch, which was responsible for investigations. Section G3 dealt with photography, chemistry, and technical research. By 1918 G3 had grown into a new H branch.

Both MI6 and MI5 had to defend the gains they made during the war in the uncertain and financially straightened years of the peace. The exploitation of wireless intercepts became a primary source of intelligence, and the codebreaking operation was concentrated in the Government Code and Cipher School (GC&CS). Bringing this operation under the control of SIS gave it a trump card in the bureaucratic infighting in Whitehall. The fact that the work of SIS had become "exceedingly technical, requiring very special qualities" in the view of the Foreign Office, meant that no other government department could present a credible claim to take it over.[21] Cumming fought his most important battle after the war was over, warding off the efforts of military intelligence to annex his service and those of government officials determined to end this waste of money and resources. In this fight for survival the support of the new secretary of state for war, Winston Churchill, was crucial. Churchill told his colleagues that with the "world in its present condition of extreme unrest and changing friendships and antagonisms . . . it is more than ever vital for us to have good and timely information."[22]

It was a warfare hitherto undreamed-of by men, a warfare at once more merciless and complicated than had ever been conceived.

W. S. Churchill, *Thoughts and Adventures*

003

The Great War and the Threat of Modernity

World War I, the Great War, is remembered as the first modern war—the conflict of machine guns and barbed wire, of poison gas and women factory workers. Its narrative is often told as a progression of deadly machines: tanks, Dreadnoughts, torpedoes, submarines, Zeppelins, and triplanes. Yet the equipment of modern warfare also included the hospital train, telegraph, camera, and tin of bully beef. Even Rolls Royce's beautiful Silver Ghost was remade into a formidable new weapon—the armored car used to great effect by T. E. Lawrence in the desert war against the Turks in Palestine. War was now fought on an industrial scale. Surveying the fifty miles behind the front, the official war historian John Buchan was struck by "the immense and complex mechanism of modern armies . . . it seemed like a gigantic business concern, a sort of magnified American 'combine.'"[1]

Not all the innovations of the First World War were mechanical or chemical; some of them merely devised new methods of fighting. The campaign in Palestine, where General Edmund Allenby was pushing the Turks back toward Damascus, was glorified in one famous book as a return to the old days of cavalry attacks and man-to-man combat. T. E. Lawrence's *The Seven Pillars of Wisdom* created the Lawrence of Arabia legend and gave tired civilian populations at home an old-fashioned adventure story to raise their spirits. Yet the war in Palestine was truly modern, hinging on air power, amphibious landings, combined operations, lightning attacks by armored vehicles, SIGINT, and aerial photography. Intelligence was paramount. Lawrence recalled that "deceptions . . . had become for Allenby a main point of strategy," and he used both the time-honored "haversack ruse," in which misleading plans were planted on a wounded or dead soldier for the enemy to discover, and the more modern creation of fake radio networks to lead the enemy into thinking that a large force was present.[2]

Allenby was open to irregular operations, and one of them brought worldwide fame to a junior member in his intelligence office. T. E. Lawrence was a graduate student in ancient archaeology when he joined Allenby's staff, and his assignment to support the Arab forces of King Faisal in their rebellion against the Turks produced some dramatic successes, a great book (some say a work of fiction), and a treatise on irregular warfare that influenced military thinkers for the rest of the century.

British armies had confronted guerilla warfare from the time of the Spanish campaign in the Napoleonic era, but the South African irregular troops that the British faced in the Boer War brought it to the fore. Winston Churchill saw the Boers' fighting potential with his own eyes while acting as a war correspondent. One of his dispatches for the *Morning Post* stated, "We are fighting a formidable and terrible adversary. . . . The individual Boer, mounted in suitable country, is worth three to five regular soldiers. The power of modern rifles is so tremendous that frontal attacks must often be repulsed. The extraordinary mobility of the enemy protects his flanks." Churchill realized that "these unpalatable truths were resented . . . but speedily vindicated by events," when 250,000 British troops struggled to contain fewer than 50,000 Boers.[3] The Boer commandos were irregulars, democratically organized and commanded, who supplied their own horses, rifles (usually bolt-action Mausers from Germany), clothing, and equipment. They received no pay, respected no authority, and as one of them recalled, "Each had his individual idea of a military salute." They traveled light, often with nothing more than their weapons, cooking tins, and blankets, living off the land and their opponents' supplies. They followed the slow-moving British columns to gather the cartridges they left in their wake. The speed and mobility of the Boers enabled them to run rings around larger forces, and in firefights their marksmanship proved superior to an army that prided itself on its accurate shooting. They exploited their knowledge of the country to stage raids into British-controlled territory, and one of them claimed, "We roamed all their territory at will."[4]

Lawrence used much the same tactics with his Arab forces during the battle for Palestine: "Armies were like plants, immobile, firm rooted, nourished through long stems to the head. We might be a vapour blowing where we listed. . . . Most wars were wars of contact. . . . Ours would be one of detachment. . . . The attack might be nominal, directed not against him, but against his stuff."[5] The stuff in question was the Turkish railway that Lawrence and his Arab irregulars continually attacked. In the greater scheme of things Lawrence's accomplishments might have been minor, but his fame was great, and he popularized the idea of well-armed individuals making daring, surprise raids. John Buchan's heroes Richard Hannay and Sandy Arbuthnot had a good measure of Lawrence in them. Churchill was especially taken with "the audacious, desperate, romantic assaults" and by the man himself, "someone strangely enfranchised, untamed, untrammeled by convention, moving independently of the ordinary currents of human action."[6] This description perfectly fits Ian Fleming's idea of a secret agent.

Warfare has always been a catalyst of innovation and new methods, and weapons were developed with blistering pace during the Great War. When Arthur Conan Doyle wrote a short story in 1913 (called "Danger") about a future war in which submarines could blockade a country, an admiral in the Royal Navy called it "most improbable and more like one of Jules Verne's stories."[7] But the submarine quickly went from a novelty to a deadly weapon. The first Holland submersible in British service, the A class of 1902, had a displacement of 110 tons and carried one torpedo tube and a crew of seven men. The E class submarines operating in the first years of the war each displaced 660 tons, carried up to five torpedo tubes, a 12-pounder gun, and a crew of 30. The German spy Klare had warned his cell mate in an English prison that "the Submarine would play a most important part in the war," and this turned out to be true; German U-boats almost starved Britain into submission, and their attacks on unarmed merchant ships helped to bring the United States into the war. This "wonderful and terrible new weapon," in Churchill's words, threatened to upend the naval balance of power and reduce the Dreadnought to an enticing target that had to be protected from submarines.[8]

While German and British navies increased the size and endurance of their submarines to make them truly oceangoing, a lesser naval power reduced the size of its underwater vessels. The Italian navy pioneered miniature submarines and human torpedoes manned by brave divers with rebreathing apparatuses and close-fitting rubber suits. In 1918 a manned torpedo (called the pig, "the Miaiale," because of its poor handling characteristics) sank an Austro-Hungarian battleship, SMS *Viribus Unitus*.[9] Instead of two great Dreadnoughts pummeling each other over miles of water, the miniature submarine brought a highly trained and well-equipped individual into close and deadly conflict with a mighty machine. Fighting underwater with equipment that looked like something out of science fiction appealed to the technological enthusiasm of many navy men, Ian Fleming among them. This form of underwater warfare was to have some significance in his war work in naval intelligence and his later literary efforts.

The most rapid (and most alarming) technological advance of the Great War took Bleriot's flimsy monoplane and transformed it into a potent offensive weapon in a mere ten years. No longer a featherweight, unreliable machine with limited range, the airplane of 1918 was capable of delivering half a ton of bombs onto targets hundreds of miles away. The first bomb delivered from the air weighed less than five pounds and was dropped by an Erich Taube monoplane (built in Austria) by the Italians on a Turkish encampment in 1911. Geoffrey de Havilland's Be 2a (*Be* stood for Bleriot experimental) was Britain's bomber when it entered the war. A single-engine, wire-and-canvas contraption, the slow and stately Be 2 could drop a 100 lb. (45 kg) bomb. Shortly after the war began, the Admiralty issued a specification for a large twin-engine aircraft, capable of carrying 600 lb. (270 kg) of bombs, with which it planned to bomb the German fleet at Kiel. The Admiralty awarded the contract to Frederick Handley-Page, and his Handley Page 0/100 was the largest bomber in the Allied fleet. With an endurance of eight hours, the 0/100 and its big brother the 0/400 formed the strategic bombing arm of the RAF, the "Independent Force" that took the war to German cities deep behind the front. Its crews called it "The Bloody Paralyser."

Many technically minded men of Churchill's generation were drawn into military aviation at the outbreak of the war. One of them admitted that the Royal Flying Corps attracted "the adventurous spirits, the devil-may-care young bloods of England, the fast livers, the furious drivers." Frederick Winterbotham joined after his cavalry regiment had their horses replaced with bicycles, and went on to fly a Nieuport biplane on the western front. Sidney Reilly enlisted in the RFC in Canada and was awarded a commission as an equipment officer. He was subsequently posted to Russia, where his official post was part of the RFC training wing.[10] William Stephenson was one of the many Canadians who served in the RFC. He had been invalided out of the infantry after being gassed, but he then volunteered for the RFC, joined 73 Squadron flying Sopwith Camels, and was an ace by the end of the war.

While the stalemate in the ground war went on for years, the fortunes of war above the trenches fluctuated dramatically with the pace of innovation. One single improvement could turn the tide of battle overnight. De Havilland's Be 2a was the mainstay of the allied air fleet at the beginning of the war, with around 3,500 of them made, but unfortunately they were obsolete before they went into production. With a top speed of 74 mph and only one .303 machine gun mounted in the cockpit, the Be 2 was no match for the Fokker "Eindecker" (monoplane), which had a forward-firing machine gun synchronized with the turning of the propeller. This single innovation established air superiority so completely that British airmen called 1915 "the Fokker Scourge" as they watched many of their comrades go to their deaths as "Fokker fodder," while their enemies referred to those flying the aircraft of the Royal Aircraft Factory as "kaltes Fleisch" (cold meat). One British pilot concluded: "ascendancy rested on the relative technical efficiency of Allied and German aircraft, and as the production of higher-performance machines passed every few months from one side to the other, and back again, so did mastery of the skies above the battlefield." Such was the outcry about the inferiority of British warplanes that Parliament carried out an inquiry in 1915 into the technological lag that cost the lives of so many Allied pilots. One member of Parliament concluded, "Our pilots are being murdered rather than killed."[11]

The Fokker Scourge started a race to build a superior fighter. Up to this point aircraft had been constructed to take off, fly in a straight line, make slow banking turns, and eventually land. This was the requirement of the army officers who directed the Royal Aircraft Factory, which was charged with the design of British military aircraft. Initially, aircraft were used only for reconnaissance, so they were designed to be slow and steady, capable of "hands-off" flying while the pilot concerned himself with map coordinates. Now they had to be reconfigured as completely different aircraft—fast, maneuverable, light on the controls, and capable of rapid, life-saving turns and steep dives in dogfights. Much of the improvement came from more powerful engines. The first combat aircraft had French or German rotary engines of around 60 hp, but soon more speed was extracted from larger 100 hp and 150 hp engines. De Havilland's DH2 entered service in 1916. Powered with a 110 hp Clerget aero engine, its top speed was around 90 mph. Along with the French Nieuport 11, with a top speed of 100 mph, it helped end the Fokker Scourge. By the spring of 1916 it was the Fokker pilots who were being killed in great numbers, and the Eindecker was removed from service as the pendulum swung back in favor of the Allies.

The single-seat, "scout" airplanes and the tactics of the dogfight came directly from the necessity of protecting the craft that photographed the enemy's lines and directed artillery fire. Successful pilots were made into celebrities, and each country's aces—Manfred von Richthofen (the "Red Baron") of Germany, Albert Ball of England, George Guynemer of France, and Eddie Rickenbacker of the United States—became national heroes and symbols of virility and courage. While the exploits of the Red Baron and the other aces kept the public's attention, the men who flew observation missions were playing the vital part of the war effort—aircraft existed only to serve the artillery that dominated the battlefields of the Great War. Von Richtofen's historic tally of 80 "kills" probably represented little more than 100 dead enemy airman, but an afternoon's work by an aerial observer might cause the death of thousands of soldiers from artillery bombardment. Aerial intelligence was an important factor in the exponential improvement of artillery's accuracy, which played a major part—along with the exhaustion of the German army—in the final breakthrough in the fall of 1918.

German aircraft manufacturers had not stood still since the success of the Eindecker, and the introduction of new Halberstadt and Albatros fighters overcame the British superiority in numbers to wreak havoc in the "Bloody April" of 1917, when the life expectancy of British pilots on the western front decreased to two weeks. In one month the RFC lost more than 250 aircraft and 400 men. Some squadrons were virtually eliminated in one week. Lt. John Slessor wrote, "I know of few things more terrifying than to be alone in a BE with an observer, and to meet seven or eight enemy scouts. One feels so utterly hopeless . . . only one gun and a maximum speed of 65 mph; whereas your Albatros has two guns and speed of 130 mph."[12] The introduction of the two-gun Sopwith Camels and SE5s brought back parity in the air war, and Fred Winterbotham was elated when his squadron got new SE5 Scouts: "these fast and reliable aircraft started freelance raids over German airspace and literally put the wind up [frightened] the German fighters."[13]

Technology Races

In January of 1915 the first Zeppelins appeared over the eastern shores of the United Kingdom. Residents reported hearing an eerie, throbbing sound above them, followed shortly afterward by the sound of explosions. The first of eight Zeppelin raids on ports along the eastern coast left only 24 people dead and 40 wounded, but the effect on morale was tremendous. Dropping a few dozen 110 pound bombs was not going to bring the country to its knees, but the Zeppelin attacks brought the war to the British homeland, and the German high command quickly escalated the bombing by attacking London. The public outcry about the Zeppelin raids forced the British government to act, enlisting men, materiel, and technology to find a way to bring the night raiders down. Armorers in service arsenals, technically minded airmen, and independent inventors pondered the problem of penetrating the Zeppelin's skin and setting fire to its combustible gas. New types of incendiary bullets were devised by serving officers John Buckingham and Frank Brock. The combination of incendiary and explosive ammunition finally made it possible to set an airship on fire, and the first Zeppelin shot down over London in September of 1916 made a national hero out of Lieutenant William Leefe Robinson, who flew a modified Be 2c up to 13,000 feet to get his shot in. Hundreds of Londoners, including the author's grandfather, rushed out to the wreck site to get a little piece of Zeppelin—mementos of the monster that had brought the war to their doorsteps.

As soon as the Zeppelin threat was countered, the Germans responded with heavy bombers, which were faster and much better defended than airships. The Gotha IVs and Vs were considered so superior to London's defenses that the first raids were made in broad daylight. More than 160 people died as the result of a raid in June of 1917, and not one enemy plane was brought down. The capital of the empire and its seat of government were shown to be defenseless, and the public was outraged. One of the onlookers of the Gotha attacks was H. G. Wells, who had anticipated such terrifying events years before in his *War in the Air.* In letters to the *Times* in 1915 he criticized the government for not deploying science in the war effort and failing to devise an airplane that could take on the marauders.

While engineers considered the problems of operating aero engines at high altitudes, lesser minds grappled with the challenge of aiming artillery at high-flying targets or designing a bullet that could penetrate the skin of a Zeppelin. More powerful aero engines meant that interceptors could fly higher to meet the threat of strategic bombers, but flying higher demanded engines that would not choke in the thin air at high altitudes and also oxygen systems to keep pilots fit enough to fight at such heights. Lack of oxygen and intense cold severely affected pilots' effectiveness and contributed to combat fatigue, which set military doctors to study the effects of high altitudes on the human body. The solutions to these problems required considerable medical expertise and the ingenuity of an amateur inventor. Frederick Sidney Cotton was born in 1894 on a cattle station in Queensland, Australia. He was educated in England, returned there to join the RNAS in 1915, and took part in night-bombing sorties over France and Germany. Promoted to flight lieutenant in June of 1917, he helped to prepare Handley Page heavy bombers for action. He noticed that the overalls worn by mechanics were much more comfortable than the standard military uniforms worn by pilots, so he went to a military outfitter and had them make up a one-piece flying suit with layers of silk and fur, covered by a light Burberry material. Sealing the cuffs at wrists and ankles kept the body heat in, and the suit came with large pockets on the thighs, which made it much easier to access maps and other gear than the slash pockets usually provided with military uniforms. The "Sidcot" flying suit was widely adopted by flyers in the Great War and used by the Royal Air Force until the end of World War II.

Innovation in the Great War spanned the ingenious inventions of individuals to the output of large, integrated technological systems, which required engineering, administrative, and educational expertise. More accurate sighting of artillery fire made antiaircraft artillery feasible, and this in turn forced the creation of command and control systems to link observers with antiaircraft batteries. It also necessitated organizational changes in the country's air defenses, which brought together different parts of the armed services under one command. The Home Guard was reorganized to provide timely air raid alerts and used audiophones, or "mechanical ears," to locate incoming bombers. The War Office established a coordinated system of air defense, and this led to a rationalization of each service's air force into one body—the RAF.

The threat of one new weapon often forced advances in other military technologies. The RNAS had developed aircraft that could take off and land on water (Churchill is credited with coining the term *seaplane*), and in 1912 Commander Charles Samson became the first pilot to take off from the deck of a moving ship. This followed the pattern of using aviation for scouting, but there was also a possibility (also anticipated by Churchill) that aircraft could bomb enemy ships. After Naval Intelligence had located the Zeppelins' base at Tondern by listening in to their radio transmissions, Sopwith Camels launched from HMS *Furious* bombed it in July of 1918—the first carrier-based air raid.

As one technical improvement led to another, armament races began among the combatants, who were continually forced to commit more technical resources to counter the latest advances of the enemy. These races occurred in every theater of the war, including secret intelligence. While German chemists devised more deadly poison gases, their Allied counterparts experimented with better gas masks; similarly, the chemists of German intelligence who were developing new secret inks were opposed by British chemists employed by Cumming to detect them. In code breaking, aircraft design, artillery ranging, and antisubmarine warfare the race was on, and the winners could alter the course of the war. While great scientists like Ernest Rutherford and famous inventors like Thomas Edison toiled in their laboratories to find a means of detecting submarines underwater, statisticians tried to find patterns in submarine attacks, and economists calculated how much food the United Kingdom needed to stay in the war. As Churchill wrote: "All the known science, every adaptation of mechanics, optics and acoustics that could play a part were pressed into service. It was a war of charts and calculations, of dials and switches, of experts who were also heroes . . . and upon the workings of this grisly process turned the history of the world."[14]

Research into theoretical physics, on one hand, and the experience gained in designing phonographs, on the other, were both important contributions to the war effort. Rutherford excused himself from a meeting on antisubmarine research in 1917 because his experiments had succeeded in splitting the nucleus of the atom. While he worked in his laboratory at Manchester University, electrical and mechanical engineers at Manchester's Municipal School of Technology developed a deep-sea hydrophone to listen to submarines, and material engineers in the Textiles Department tested new fabrics to be used in aircraft construction. Some of the small but vital innovations in air warfare were made by mechanics and riggers working in the machine shops of the RFC and by the pilots themselves as they redesigned, disabled, or added new pieces of equipment to their planes, some of it captured from the enemy. Although Anthony Fokker is credited with producing the interrupter gear that enabled a machine gun to fire through the Eindecker's propeller, similar devices were produced by officers and mechanics all along the front. The Alkan interrupter was the work of Sergeant Mechanic Alkan and a navy engineer. The difficulties of handling the 18 or 24 glass negatives used by cameras for aerial photography was alleviated by a cassette system devised by the German pilot Fritz Dubowsky.[15] These innovations were the work of individuals and were usually named after them, such as the "Anderson Arch," a flexible mount for machine guns.

World War I brought innovation firmly into modern warfare and established a precedent for governments to mobilize scientific and engineering expertise. In 1909 Secretary of State for War Richard Haldane created the Air Advisory Committee to investigate the new science of aeronautics. In 1915 the Board of Invention and Research (BIR) was established to solicit expert scientific assistance to solve technical problems. A "Physics Department" was set up at the Royal Aircraft Factory in Farnborough to carry out research. It employed university professors, Fellows of the Royal Society, and several scientists who became Nobel Laureates. One of the civilian scientists who worked there was Frederick Lindemann, a bright young man educated in a German Technische Hochschule who had been a research assistant to the great German physicist Walther Nernst. At Farnborough Lindemann worked on methods of detecting aircraft by sound and infrared rays, and he designed instruments like turn indicators, bomb sights, and range finders. On the roof of the Physics Department a young meteorologist recorded weather data and produced a daily forecast. Robert-Watson-Watt also experimented with radio waves as a means to detect the approach of thunderstorms. Henry Tizard graduated with a double First in mathematics and chemistry from Oxford, and had also worked for Professor Nernst in Berlin before he joined the Faraday Laboratory of the Royal Institution. He enlisted in the RFC and joined the Central Flying School as a scientific experimental officer. He carried out research on bomb sights, aerial cameras, and radio equipment. In 1918 he became an administrator of the newly created department of Research and Experiments at the Air Ministry. After the war he joined the Department of Scientific and Industrial Research, which had been established in 1916 to mobilize science and technology for the war effort. Archibald Vivian Hill had worked in Cambridge University's Physiological Laboratory before the war and helped found the new field of biophysics. In 1915 he was made head of the Anti-Aircraft Experimental Station and used mathematics and operational research to assess the different methods of ranging antiaircraft artillery.

At war's end each of the armed services had set up departments of scientific research. Heading the Air Ministry's department was Henry Wimperis, a university-trained engineer who spent the war in the RNAS. He participated in the research to detect enemy aircraft and improve the RAF's navigational and bombing aids. The RAF's Mechanical Warfare Experimental Establishment was testing new forms of propulsion, new theories about aerodynamics, and new designs of aircraft and bombs. The Admiralty continued research in submarine detection, and the army experimented with new techniques of gunnery and signaling. A chemical warfare experimental station was set up at Porton Down to experiment on chlorine, phosgene, and mustard gas. By the 1920s there were 60 scientists working there.[16]

Technological Anxieties

The technological advances of the Great War made the outlook for peace in the postwar world uncertain. By 1918 the British had a massive four-engine bomber, the Handley Page V/1500, which was designed to bomb Berlin, and the Germans had the prototype of a gigantic 10-engine triplane that (they hoped) could reach New York. The Swedish naval attaché in Berlin wrote, "Compared to this giant the latest enemy airplanes used in the attacks against London are merely toys."[17] Aerial reconnaissance had begun in 1914 with sketch pads and perhaps a personal camera taken aloft, but by the end of the war German aircrews with oxygen equipment were using electrically operated cameras in high-flying Rumpler C.VII reconnaissance planes, whose supercharged engines could reach 24,000 feet.[18] By 1925 the RAF was employing the F-24 camera, which used a roll of film to take up to 125 images. An airplane that could fly at 100 mph would have been considered virtually unattainable before the war, but top speeds edged closer to 125 mph by the end of the conflict. One of the requirements of the armistice of 1919 was that the Germans had to hand over all models of their latest fighter, the deadly Fokker D VII, to the Allies. Some enthusiastic young technocrats were anticipating that the next war would be fought entirely in the air.

The Great War had for the first time brought terror to entire populations. Powerful new weapons, which often came as a nasty surprise, could significantly affect the outcome of the war, if not by material damage then by terror. Scholarly analysis after the war found that the threat of "Zeppelinitus" had been greatly exaggerated, as had the damage done by the Fokker Scourge, but this mattered little in the minds of the public and did nothing to allay the fears that forced governments to act.[19] Much of the terror of night bombing was in the spectacle rather than in actual physical damage. A house in Stoke Newington was the first in London to be hit by an incendiary bomb. Aviation historian Ian Castle imagined the scene: "There's never been anything like this. Suddenly a blazing bomb is coming out of the sky and setting light to a house, it's almost science fiction." In twenty minutes a Zeppelin dropped 3,000 pounds of bombs on London and left seven people dead, one of them a three-year-old girl who died in her bed. After that the British called Zeppelins "baby killers." Doris Cobban was five years old when Zeppelins appeared over the London suburb of Lewisham: "I remember my father coming up to the bedroom and he picked me up and wrapped me in a blanket and he said this is history, you must see this."[20] The Zeppelins came out of the night sky, invisible, their presence revealed by the throbbing of their Maybach diesel engines. As they came into view and searchlights from the ground picked them up, their ominous shape and gigantic size made for unworldly images. Despite the danger, thousands of Londoners went out on the streets to watch the Zeppelins make history and produce a spectacle that left an indelible memory. One of those onlookers was the young Alfred Hitchcock: "The whole house was in [an] uproar. . . . Outside the window shrapnel was bursting around a search-lit Zeppelin—extraordinary image!" He told a French interviewer that this was the first time he experienced genuine fear.[21] His film *Secret Agent* (1936) begins with a scene in which "R," the head of the Secret Service, briefs one of his agents while through the window of his office we see searchlights lighting up the sky above London, seeking out the ominous Zeppelins. Another young boy who remembered the air raids was R. V. Jones. His early years were steeped in "the experiences of the air raids on London. . . . The shattered houses that I saw then, and the suspense of waiting for the next bomb, remained in my memory as the Second World War approached."[22]

On 8 September 1916 a single Zeppelin out of a large raiding force reached the capital and did spectacular damage, starting fires, killing more than 20 people, and injuring nearly 100: "London lay, breathing heavily, oppressed by a nightmare such as the most ferocious minds of the darkest of the Dark Ages did not dream of . . . terrifying some of us beyond mental endurance; making us all, strong and weak, profoundly wretched and uneasy, filled with a restlessness that was worse than pain."[23] The damage to morale was recognized as an important part of this strategic bombing campaign. Peter Strasser of the German navy's airship fleet claimed: "We will force a black out of industrial plants. . . . We could become a strong factor both psychologically and materially in weakening the Allies in their own backyard." German intelligence was focused on assessing the damage to morale, and the high command obtained a secret Admiralty report that lamented the lack of an effective response. The *Kolnische Zeitung* newspaper crowed: "The most modern air weapon, a triumph of German inventiveness and the sole possession of the German military, has shown itself capable of crossing the sea and carrying the war right to the sod of old England."[24] No longer safe as a small island protected by a large navy, the United Kingdom looked anxiously skyward, and as one civilian lamented: "The raiders have London at their mercy. I kept saying to myself, there are no defences against them." The largest city in the world, the capital of the empire, was, as Churchill pointed out, "the greatest target in the world . . . a fat cow tied up to attract beasts of prey."[25]

As Britons crouched in cellars during these air raids, they had plenty of time to consider the march of events and machines that had brought them to their present moment. The technological enthusiasm of the Edwardian era came to grief during World War I, giving way to dystopian anxieties in the 1920s and 1930s. Those terrified civilians described by Violet Hunt and Ford Madox Hueffer "talked about Arkwright of the spinning Jenny, and Stephenson of the Rocket, and the other Stevenson of the lighthouses, and Caxton who printed things. . . . And a pretty mess they've made of it all! . . . What it all led to . . . the shadowy London that we know, to these Zeppelin nights; to this immense strain; to all the tears."[26]

Death Descends on the City.
Metropolis

004 Imagining the Future: Technology on Film

Photography was one of the wonders of the Second Industrial Revolution, and cameras were prized by technological enthusiasts. When Samuel Langley of the Smithsonian Institution launched his glider *Aerodrome 6* from a boat on the Potomac in 1896, Alexander Graham Bell stood on the bank with a camera in hand.[1] The Wright brothers were enthusiastic photographers and developed prints in their own darkroom. On that historic day in December 1903, Orville Wright set up a camera at the end of the launching track at Kitty Hawk and asked a local man who had been helping them to take a picture when the machine took off. John Daniels caught the Flyer at the moment it lifted off the track, producing one of the iconic images of the twentieth century. Life was speeding up at the turn of the century, and the motion picture camera was there to record it. The first short films were "actualities," as pioneers like the Lumières turned their cameras on the world around them, often reflecting their technological enthusiasm in filming locomotives, steamships, and motorcars. But as filmmakers grasped the potential of film cameras to recreate rather than record, Edwardian enthusiasts like Georges Méliès used them to imagine the future. While Jules Verne's *Around the World in Eighty Days* recorded the transportation revolution of the Victorian era in 1873, Méliès's *Impossible Voyage* introduced airships and submarines to audiences in 1904. Verne's best-selling books were popular with filmmakers. *From the Earth to the Moon* inspired Méliès to make *A Trip to the Moon* (1902), and several versions of *20,000 Leagues under the Sea* were produced: American Biograph's 1905 short, then Méliès's in 1907, and finally Universal Studios' full-length feature in 1914.

The art of Verne and Wells was convincing enough to influence life. The German technological enthusiast Hermann Oberth (born 1894), who is credited with being one of the fathers of space travel, said that he had read Jules Verne's

From the Earth to the Moon so many times that he knew it by heart. Oberth's book *By Rocket into Interplanetary Space* (1923) is considered, with Robert Goddard's "A Method of Reaching Extreme Altitudes," one of the foundational texts of space travel. Goddard, who built the first successful liquid-fueled rockets in the 1920s, acknowledged the influence of Wells's *First Men in the Moon* and *The War of the Worlds.* The popular books of Oberth, Willy Ley, and Max Valier promoted the feasibility of interplanetary space travel, and Ley and Valier were founding members of the Verein für Raumschiffahrt (VfR, Spaceflight Society) in 1927, which brought together many German spaceflight enthusiasts, including the young Wernher von Braun. Rocket ships and planes captured the imagination of Weimar Germany so much so that Michael Neufeld has described a "rocketry and spaceflight fad" centered on Berlin from 1923 to 1933. Enthusiasm about space travel combined with fantastic images from films bolstered optimism about technological progress and the role of Germany as a leader in modernity.[2]

One ambitious German film was to have great influence on space travel. Fritz Lang's *Frau im Mond* (*Woman in the Moon*) grew out of his friendship with Willy Ley, who called on his friend Oberth to join the production as "Scientific Collaborator." Oberth helped with the design of the moonscape and the film's spaceship, providing technical details that gave the film an air of reality. One intertitle informs us that "the period of acceleration . . . can be deadly for the human organism if it is greater than forty meters per second." Much of *Woman in the Moon* is pure fantasy, such as the breathable atmosphere on the moon, yet it does have several technical details that anticipated future space technology. In the film the giant multistage rocket stands upright and is moved on rails to the launch point—the same way that NASA does it today. We see astronauts experiencing weightlessness in flight, and their ship follows a trajectory that bears some resemblance to the flight plans of future moonshots. The most prescient feature of the film is the backward countdown that precedes liftoff: the moment of highest drama for every rocket launch since 1929 is the creation of a silent film!

As part of the marketing of the film, Lang had Oberth build a working rocket in the Ufa studio workshops. The rocket was to be launched when the film premiered.[3] Although Lang and others made claims that this rocket was the prototype for liquid-fueled technology, it is unlikely that Oberth and his collaborator Rudolf Nebel, a World War I flyer, had the expertise or materiel to make a working rocket. Nevertheless, this piece of set design was to have a long life in the promotion of space flight and the VfR. Lang always claimed that this model spacecraft was so scientifically accurate that the Nazi Party later confiscated it in order to keep this important military technology in Germany. Although no more than a prop, this creation inspired many young men to embrace rocketry and space travel, Werner von Braun among them.[4] Von Braun probably joined the VfR as a result of watching *Frau im Mond* and could have helped Nebel in the construction of the rocket. It is probably no coincidence that the German army began to support experiments in rocketry in the same year that *Frau im Mond* came out. Both von Braun and Oberth joined the Nazi Party to get official support for research into rocketry, and von Braun went on to lead the team of scientists and engineers, which included Oberth, who developed the V-2 ballistic rocket at Peenemunde during World War II. When the war was over, von Braun went to work for NASA in Huntsville, Alabama, and was soon joined there by Oberth, who left his job with the Italian navy to work with his old boss. Fritz Lang was invited to give a talk to Wernher von Braun's NASA team in Huntsville in 1968, and he recalled that they "considered me in a certain way as the father of rocket science."[5]

Filmmakers were quick to show the horrors of modern technology. H. G. Wells had anticipated the threat from the skies in *The War in the Air* as German airships devastate New York—"the first of the great cities of the Scientific Age to suffer by the enormous powers and grotesque limitations of aerial warfare."[6] This new form of barbarism was depicted in *The Airship; or, A Hundred Years Hence* (1908), *The Airship Destroyer; or, Battle in the Clouds* (1909), and *Aerial Anarchists* (1911), which was based on Verne's 1886 book *Robur le Conquérant. The Airship Destroyer* imagined an air raid in which artillery shells are dropped from airships and provided a prescient vision of aerial combat between airships and airplanes. Its intertitles summarized one of the popular plots of science fiction: "Defense: The inventor with the assistance of his sweetheart sends his airship destroyer on its mission of vengeance. The torpedo, steered through the air by wireless telegraphy. One flash and the airship is doomed."

Real images of the horrors of war were captured by a few intrepid cameramen who ventured close to the front during the Great War, and melodramatic recreations were produced by film studios. *The Little American*, a Cecil B. DeMille production of 1917, catalogued the atrocities facing civilians as the heroine (played by Mary Pickford) suffers at the hands of German submarines and bombers before being saved from a German firing squad after being accused of spying. The horrors of the Great War influenced postwar filmmakers who envisaged an unsettling new modernity fermented by industrial technology.

Fritz Lang's *Metropolis* (1927) presented an unforgettable vision of a future run by machines that enslave the lumpen proletariat. *Metropolis*'s vertical cities, inspired by Lang's visit to Manhattan, house a technical elite led by Joh

Fredersen, who commands the labor of a downtrodden working class. The machine halls in which the workers toil look like power stations with their massive steam-fired machines, giant crankshafts, and control boards of dials and switches. In one of the film's most expressive scenes this giant machine turns into the mythic monster god Moloch, who devours the workers. If there is a villain in *Metropolis,* it is the inventor Rotwang, a pioneer in the long line of mad scientists portrayed in silent films. Lang's Rotwang wears a crumpled lab coat, and his disheveled look adds to the impression that he might not be sane. He creates a robot in a series of striking images—currents of electricity arc through the air, and liquids bubble in laboratory glasses—as he turns the dials and levers in his laboratory. In a scene prescient of James Whale's *Frankenstein* (1931), Rotwang kidnaps the saintly young girl Maria and connects her to his equipment to transfer her life force to his creation. Rotwang's robot is a doppelganger for Maria the good, but the robotic Maria is a subversive force that encourages the proletariat to rebel and destroy the city. Tom Gunning makes the point that the character of Maria represents the two sides of technology: one an obedient servant of humankind, the other an uncontrollable destructive force, just as Fredersen represents the rational technocrat and Rotwang the demonic inventor in a Manichean division of good and evil.[7]

This dichotomy between utopian and dystopian technology is revealed in the set, which contrasts the dark and cluttered laboratory of the inventor with the airy conference rooms of the technocrat filled with modern furniture and the latest office equipment. Both are seats of power—one uncontrollable and malignant, the other an example of modern scientific management. Far above the massive machine shops and Rotwang's underground laboratory, Fredersen runs the city with modern office equipment like telephones and television screens. Electrical annunciator boards with flashing lights and switches on his futuristic desk keep him in control. He can see all parts of the city via screens in his office and can communicate with all his workers by telephone.

The villain as a man of science can be traced back to Sherlock Holmes's greatest nemesis Professor Moriarty. A brilliant mind, a professor of mathematics who wrote a treatise on the Binomial Theorem as a youth, Moriarty runs a criminal empire with the efficiency and ruthlessness of an organized-crime syndicate. As one of the first master criminals in fiction, his devious plots to murder, blackmail, and rob come from a brilliant but antisocial mind. Mary Shelley's *Frankenstein* (1816) provided filmmakers with another vision of the mad scientist. The Edison Company made the first filmed version of *Frankenstein* in 1910, and this was followed by a stream of silent and then sound films culminating in the

Universal Studios version in 1931, which was so successful that it spawned a series of sequels. In *Frau im Mond* we see that scientific genius often verges on madness. Professor Manfeldt (Klaus Pohl) has been laughed out of the scientific community because of his wild predictions about space flight, but when his spaceship lands on the moon, his vindication is marred when he overrides the control system to leave the spacecraft and seek out water on the surface of the moon with a divining rod.

Future Wars

The hangover created by the terror weapons of the Great War continued into the 1920s and 1930s in books, films, and comics that predicted a dangerous future of secret weapons and alien invasions. H. G. Wells's *War of the Worlds* (1897) was the mother of all paranoid invasion stories. The threat in this case came not from Germany but from Mars, but the book sounded the same alarm about apathy to a technological threat, which would endure in English literature up to the time of Fleming and beyond. The story starts with a "a colossal puff of flame" from space, "yet the next day there was nothing of this in the papers . . . and the world went in ignorance of one of the gravest dangers that ever threatened the human race."[8] A fully automated enemy with advanced weapons such as "the heat ray" easily overcomes all the forces of the government. Wells's heat ray anticipated the death rays that populated science fiction and occupied military research laboratories for the next hundred years.

In his 1940 essay on boys' adventure books, George Orwell traced it from the school-based stories of the prewar decade to the newer publications of the 1920s, which offered exciting tales of the Wild West, Frozen North, crime, and the Great War. Orwell pointed out: "The one theme that is really new is the scientific one. Death rays, Martians, invisible men, helicopters and interplanetary rockets."[9] The reality of a death ray emerged in the 1920s when several individuals claimed to have developed high-energy particle or electromagnetic beams that were strong enough to cause injury. These included well-known scientists like Guglielmo Marconi and Nikola Tesla, along with enthusiastic amateur inventors Harry Grindell Matthews and Edwin R. Scott. Many of these rays were intended to defend against aircraft. Grindell Matthews claimed to have received financing from the British government during the war to invent a beam: "as it touches the plane it bursts into flame and crashes to earth." He later developed this into what he called a "diabolical ray." Tesla claimed to have invented a beam "of such tremendous energy that it will bring down a fleet of 10,000 enemy airplanes at a distance of 250 miles."[10] The death ray was firmly established in popular culture by the 1930s and was often the weapon of choice (with poison gas a second) of the villain in science fiction serials and films. Death rays took center stage in *The Vanishing Shadow* (1934), *The Phantom Empire* (1935), and *The Lost City* (1935).

The makers of the hugely popular American sci-fi serials of the 1930s maintained a dystopian outlook by depicting a Wellsian view of world wars, mad scientists, and death rays. Science fiction heroes like Flash Gordon and Buck Rogers did their swashbuckling in outer space and faced a variety of high-tech threats coming from the skies. They faced alien invasions, powerful projectiles, biological and psychological warfare, robots, deadly rays, and the malicious intent of foreign or extraterrestrial villains. Much of the villains' threat comes from their (almost) unimaginable high-tech weapons: "This machinery is nothing like I ever saw before," says a character in a Buck Rogers adventure. The bad guys' goal of world domination is attempted through scientific research, "the latest fiendish invention," which reflects contemporary fears about technological advance. Buck Rogers confronts "mysterious machines from outer space" that are radio-controlled; Flash Gordon has to contend with delayed-action mines; and Ace Drummond fights off deadly radio waves. Emperor Ming of the Flash Gordon serials comes up with poison gas, "death dust" that leads to "the purple plague," zombies, robots, "walking bombs," death rays, and explosive projectiles over the course of his career. Many reincarnations of these weapons can be found in James Bond films. The giant rock drill or circular saw of early science fiction, which bears down on the helpless hero tied down in its path, becomes a "death ray" creeping closer to the tightly bound Dr. Zarkov in "The Destroying Ray" episode of Universal's 1936 *Flash Gordon* serial. The same scene is duplicated in *Goldfinger*, and in this case it is a laser beam bearing down on James Bond in the film's most memorable sequence.

Just like the Bond stories, the plots of the sci-fi serials depend on one technically savvy and well-equipped hero to save the world. Ace Drummond, Buck Rogers, and Flash Gordon all represent the hope of technological enthusiasm. All are pilots at a time when aviation represented the high point of science and technology. The character of Ace Drummond was created by Eddie Rickenbacker, "America's Beloved Ace of Aces" in the Great War and "The Inspiration of Youthful Airmen the World Over." These technological enthusiasts are comfortable with machines and race around the plots in airplanes, spaceships, sports cars, and motorbikes. Their costumes reflect the dress code of pilots, with leather flying helmets, goggles, and one-piece Sidcot flight suits. They face technological threats with judicious use of violence and technologies of their own: televisions, atomic shells, helmet radios, special protective suits, and a wide assortment of ray guns—many can blow things up, others cool things down, and some can even make the bad guy disappear. Without these gadgets the hero is powerless, and somehow he always comes up with a technological solution to every threat—the trick is discovering it before the episode runs out.

Scientific Intelligence

Although Buck Rogers and Flash Gordon could depend on an armory of world-saving technology, ordinary humans could only hope that a solution could be found to the threat of the long-range bomber. This became a matter of pressing concern in the 1920s and 1930s as the public watched longer and longer flights, such as Lindbergh's crossing of the Atlantic in 1927. While advocates of strategic bombing promoted the idea of air power in the 1930s, its potential victims watched films like *Things to Come* (1936), which was based on a 1933 H. G. Wells story *The Shape of Things to Come*. The book anticipated the social and political development of Europe up to 2016 and predicted a Great Depression and a general war that was to begin in 1940. *Things to Come* presented many prescient machines—helicopters, wrist radios, streamlined tanks, and poison gas—but at its heart is the potency of a strategic bomber that destroys cities with massive bombing raids just as H. G. Wells had predicted.

The film was a British production, and it was made during a period of intense anxiety over the vulnerability of the United Kingdom to air attack. Winston Churchill had read Wells's *The War in the Air* "with astonishment and delight" and loudly advocated building up Britain's air defenses. He had also digested the theories of General Giulio Douhet, who predicted that air attacks on "vital centers" and the enemy's morale could end a war in weeks rather than years. Churchill learned an important lesson from the Great War: "the possession by one side of some overwhelming scientific advantage could lead to the complete enslavement of the unwary party . . . reducing their opponents to absolute helplessness."[11] During the 1930s, British government estimates of the casualties of an air attack on London ranged from 28,000 to 40,000 killed or injured. Rearmament became a major political issue, and only a few weeks after Hitler came to power in 1933, the British announced an increase in the size of the RAF, setting a target of 1,250 fighters to defend the United Kingdom. These air estimates became a political issue in the mid-1930s, and the threat of enemy air power generated much public concern. With the benefit of hindsight historian David Edgerton has demonstrated the effectiveness of the Air Lobby, the British aviation industry's political arm, which constantly overestimated the power of the weapons the government was manufacturing and bloated the number of casualties a bombing raid would produce. Churchill later admitted that he exaggerated British weakness against aerial attack to "act as a spur" to further develop its defenses.[12] How much weight science fiction played in building fears of future wars is impossible to know, but films like *Things to Come* provided a graphic illustration of what happens to a country that does not build enough aircraft to protect its airspace. Prime Minister Stanley Baldwin's famous conclusion in 1932 that "the bomber will always get through" was supported by politicians, scientists, military leaders, and filmmakers. Lindemann wrote a letter to the *Times* in 1934 asserting that "there is at present no means of preventing hostile bombers from depositing their load of explosive, incendiary materials, gases, or bacteria upon their objectives." World War I veteran and future prime minister Harold Macmillan looked back on those anxious days of the mid-1930s and remembered: "we thought of air warfare . . . rather as people think of nuclear warfare today.[13]

The threat of German air power was real. Although it had been denied an air force by the Treaty of Versailles, it was known in the intelligence community that German pilots were being trained in secret and that airplanes currently being built as civil airliners were intended to be used as bombers. When Adolf Hitler announced the formation of the Luftwaffe in 1935, it had already been in existence for some time and allegedly numbered more than 1,800 aircraft. Comparisons between the numbers of RAF versus Luftwaffe aircraft became a hot-button topic in the United Kingdom, with newspapers, aviation experts, and politicians weighing in. A scene in *Things to Come* shows the newspapers proclaiming, "Another 10,000 Fighters!" Consequently one of the primary goals of British intelligence was to establish the exact number of aircraft available to the Luftwaffe. Right after the war, SIS added a fourth section to its organization specifically charged with keeping up with rapidly advancing aviation technology. In 1930 Frederick Winterbotham was recruited to join the staff of the Royal Air Force, where he was assigned to this department. During the next few years he built up a technical intelligence network for the RAF to gather information on the development of military aviation in hostile or potentially hostile countries. He was charged with getting information about new aircraft like the Heinkel 111 bomber, and "aircraft armament and performance and the possible production of any new type of gun or bomb." Winterbotham was not a desk-bound bureaucrat but an adventurer who did some spying of his own on frequent trips to Germany before the war, where he met with Adolf Hitler and built up friendships with many influential figures in the Luftwaffe. Winterbotham recruited agents, analyzed their information, and wrote reports for the air staff on the numbers of German aircraft and their capabilities. Of primary concern to the air ministry was the development of new secret weapons. Winterbotham investigated rumors that a radical new type of aero engine was being developed that would increase speeds to unimaginable levels. He said of his work: "As always I would be listening for any hint that a new secret weapon had been invented, for news of German radar research, and for any suggestion of advances in their development of aerial photography and code techniques."[14]

The threat of air attack pushed the government to mobilize its scientific resources, and in 1934 the Air Ministry set up the Committee for the Scientific Survey of Air Defence (CSSAD) to consider how advances in scientific and technical knowledge could be used to defend the country against attacks from the air. The committee brought together several prominent scientists, including A. V. Hill, Patrick Blackett, and Frederick Lindemann, and was chaired by Henry Tizard, who was able to persuade the Air Ministry to appoint a director of scientific research for the air force (H. E. Wimperis). The outlook was not reassuring. In June of 1934 A. P. Rowe, the secretary of CSSAD concluded that "unless science evolved some new method of aiding our defense, we were likely to lose the next war if it started within ten years."[15] Tizard also pushed for the hiring of a scientific intelligence officer to liaise with Winterbotham's department. He came to this decision after a meeting with Winterbotham and Robert Watson-Watt to discuss what to make of a report of a strange object, a "saucer with a central aerial" that had been located by agents on the Baltic coast. After the meeting Tizard decided to provide Winterbotham with a scientist to assist him in "this sort of scientific work." The young, academically trained physicist R. V. Jones was made scientific officer to the Air Ministry.

When Jones took up his position, he found that the concern about the threat of secret weapons was so great that he was allowed to look through all the files held by SIS that might give some clue about German scientific activities. This gave a civilian unprecedented access to the nation's secrets and showed how seriously Whitehall was taking hostile science. Jones found a long list of claims of wonderful new weapons, and he looked through the files on eccentric inventors like Harry Grindell Matthews, but there was no sign that any of these experimental projects would eventually bear fruit. Nevertheless, the threat of secret weapons of unknown potency loomed large in the minds of the government and public. Science fiction had become political reality, and the Wellsian vision of a world destroyed from the air became a public issue. When the real war started in 1940, Tizard reported: "We have no reason to be particularly alarmed that the enemy is substantially ahead of us, on the whole, in the technical development of known weapons and equipment. Both sides are however frightened that unknown weapons might be sprung on them."[16]

Nazi Germany was rumored to have a wide variety of secret weapons in development, including a "death ray" that used radio waves and was capable of destroying cities and people. On the eve of war in Europe Archie Boyle, the director of intelligence at the Air Ministry, came to the pessimistic conclusion that "our arrangements for the study of scientific intelligence are inadequate."[17] He and many of his colleagues were alarmed at continual reports of German secret weapons. The German chancellor, Adolf Hitler, had recently made a speech referring to a weapon against which there was no defense, which sent military men to the Secret Intelligence Service clamoring for evidence of "death ray machines" and the like. This turned out to be no more than a faulty translation of the Führer's speech but did little to calm fears that the Germans had something nasty up their sleeves. As part of the investigations of this committee, Wimperis asked Watson-Watt, director of the Radio Research Station, about the possibility of building a British death ray that could be used against invading aircraft. Watson-Watt quickly returned a calculation carried out by his colleague Arnold Wilkins showing that the power requirement of such a device was impossibly large; thus, the fears of a Nazi death ray created in the spirit of science fiction serials came to an end. Moving from the impossible to the improbable, Watson-Watt mentioned in the same report a suggestion that was originally made to him by Wilkins, who had recently heard of aircraft disturbing shortwave communications as they flew through the signals. Watson-Watt suggested that radio waves might be capable of detecting aircraft. The government bureaucracy moved remarkably quickly, and Tizard immediately presented the idea to CSSAD. Within a month Watson-Watt sent the secret memo "Detection and Location of Aircraft by Radio Methods" to the Air Ministry, and tests carried out at the BBC's shortwave broadcast station at Daventry proved that radio waves could be reflected by an aircraft. On 2 April 1935 Watson-Watt received a patent on a radio device for detecting and locating aircraft. This episode revealed that science fiction could play a useful part in creating science fact and added another chapter to the technological "silver bullet" narratives in comics, radio programs, and films of the 1930s.

In the 1930s historians of technology like Lewis Mumford looked for salvation in the goodwill of the machine makers and systems builders who were going to create a more rational and humane world through enlightened engineering and regional planning, while popular culture often found villains in the technological elite. In the sci-fi serials every mad scientist was thwarted by a hero who found a savior in a benign technology or scientific helper: Flash Gordon had Dr. Zarkov, and Buck Rodgers was assisted by Dr. Huer, and these are the men who usually come up with antidotes to the evil machinations of the villains. Some real-life scientists at Cambridge University founded a scientists antiwar group in the mid-1930s who staged demonstrations objecting to the use of science for the cause of war, but pacifism was always a last resort in popular culture. *Things to Come* depicted a civilized world taken back to a medieval society, but redemption was found in the "old engineers and mechanics . . . the last trustees of civilization when everything has failed," who build a new and better world. But this technological elite enforced good behavior with strategic bombers that drop a debilitating "gas of peace" on uncivilized miscreants. Although the brigand chief at the receiving end of this form of advanced warfare laments that "science is the enemy of everything," the film demonstrates that it can bring peace, prosperity, and a contented generation of techno-savvy consumers in postwar society. As a little girl tells one of the brotherhood of engineers and mechanics: "They keep on inventing things now, don't they, making the world lovelier and lovelier."

> The hero equipped with powers superior to those of a common man has been a constant of the popular imagination—from Hercules to Siegfried . . . all the way to Peter Pan.
>
> Umberto Eco, "The Myth of Superman"

005 Spy Films

Spies and secret agents were common characters in silent films, in which they often acted just like detectives in an urban drama or soldiers fighting behind enemy lines. Disguise, deceit, and surveillance play their part, and the audience gets the added satisfaction of knowing things that the characters don't; they carry out their own surveillance on the actors. The spy film also gave opportunities to exploit exotic locations and create heroic characters. The 1910 Edison film *Michael Strogoff* is an early example: a story about a soldier of the Russian Imperial Guard who is entrusted as a courier to go into enemy territory. The film has scenes of a romantic train journey, the hero's imprisonment and interrogation, and finally a climactic duel with the villain. Strogoff is a man of action, endlessly resourceful and attractive to the opposite sex—making him one of the earliest incarnations of James Bond. Crime films had already depicted much of the work done by spies, and they provided the model for the action scenes of espionage films: car chases, burglaries, surprise attacks, safecrackings, escapes, explosions, and interrogations.

Crime dramas also enhanced the character of the villain in popular culture. Fleming's generation grew up with bad guys who often represent the Other, such as demonic and deceitful continentals like Fu Manchu or Ming the Merciless. These villains made a big impression on Fleming in his early "Doctor Fu Manchu days . . . the adventure books of one's youth."[1] They came in a variety of dark shades and foreign accents and wore outlandish clothing. They are stern, cruel, and loquacious, and their weakness is normally white women. Like the evil masterminds that James Bond has to face, they are highly intelligent men who use science for their own evil ends, which is usually world domination. They enjoy torture and cruelty, and they indulge in long dialogues with the condemned man: "You will work for my conquest of the universe" or face "an amusing death which

I have devised," Ming the Merciless tells his helpless captive in episode 4 of *Flash Gordon Conquers the Universe* (1940), anticipating the obligatory exchange between the captive Bond and the taunting villain.

Louis Feuillade's *Fantômas* series of dramas (1913–14) established the character of the criminal mastermind who employs technology in the execution of sensational crimes. Filmmakers like Fritz Lang acknowledged the debt they owed to Louis Feuillade, and the link between his Fantômas character and Lang's Dr. Mabuse is clear. Mabuse is a scientist as well as a criminal, who uses hypnosis to control his victims and has cars fitted with poison gas chambers to get rid of them. Both are masters of technology in the service of evil; Fantômas uses a dead man's hand as a glove to conceal his real fingerprints and detonates a building (with his pursuers inside) with a flick of a switch. Feuillade's films were popular on both sides of the Atlantic, and his *Fantômas* series had much in common with the Bond franchise; both had recurring characters based on best-selling fiction, and these larger-than-life heroes and villains made the actors who played them celebrities. "Fantomania" broke out among fans as the films reached a wide audience, just as "Bondmania" did 50 years later. The pieces of action invented by these pioneer filmmakers had long lives. Two scenes that appear in *Juve vs. Fantômas* (1913) were remodeled 50 years later in *Dr. No:* a snake enters the hero's bedroom at night in an assassination attempt, and a character uses a breathing tube to stay underwater and escape detection.

Spying brought drama to the film narrative with its scenes of violence and the transformative effect of disguise, but other than hiding or stealing secret documents, spy craft was not important to the plot. What attracted novelists and filmmakers to the spy story were the colorful characters, such as Mata Hari—the exotic dancer shot as a German spy in the Great War. *Mata Hari,* a 1927 German production, and the very popular *Mata Hari* of 1931, which starred Greta Garbo, established the character of the seductress/spy that was to become commonplace in spy fiction. Mata Hari might have died in 1917, but the fusing of sexual promiscuity with espionage was such a popular combination that it dominated spy fiction from then onward. Spy films usually brought drama and sexuality together, but one unusual silent comedy proved that this union had more than one combination. *Yankee Doodle in Berlin* (1919) was a Mack Sennett comedy. It starred the female impersonator Bothwell Browne, who plays the role of Bob White, "US Aviator" and thus a technological enthusiast. White lands his plane behind enemy lines and inveigles his way into the Kaiser's inner circle by doing what he does best—impersonating a woman. Although played for laughs, *Yankee Doodle in Berlin* was no one-reel, slapstick comedy but a full-length feature that examined the work of espionage in some detail. Like many agents who were later dropped into Occupied Europe in World War II, Bob White was accompanied by his own radio operator and was equipped with delayed-action explosive devices. He/she operated in disguise and was pretty handy with his fists in resisting arrest by German guards (rather than "unarmed combat," this violent behavior was called "Football tactics" in the intertitle). *Yankee Doodle in Berlin* was a piece of propaganda that made relentless fun of the Germans, but it had an impressive amount of content about the recent air war, bolstered by stock footage of US warplanes. Called "the messengers of democracy," squadrons of bombers wreak havoc on the Germans below. Here for the first time a film fused the terror from the sky with the work of espionage.

The father of the spy film was Fritz Lang. His *Spione* (Spies) of 1928 introduces us to a dashing secret agent who is known only by his number, as well as a master spy whose goal is to take over the world. What links *Metropolis* with *Spione* is the way that characters operate within a technological web that enables them to run large, complex organizations. The spymaster Haghi runs his criminal network from a Bauhaus-style desk, fitted with pushbuttons, telephones, intercoms, and pneumatic tubes. To play Haghi, "the most dangerous man in Europe," Lang cast Rudolf Klein-Rogge—the same actor who would play the master villain Dr. Mabuse in Lang's trilogy of crime dramas. Haghi hides behind a respectable facade in an underground maze of offices "and rooms equipped with countless sending and receiving apparatus." The hero of *Spione* is Agent 326 (played by Willy Fritsch): handsome, faultlessly dressed, who lives in a fancy hotel with his manservant, drives a very nice car, and likes to travel by international train and by air. His boss, the head of the Secret Service, acts as a sort of father figure and depends on Agent 326 to save the day and end the helplessness of the service. The film shows us the essential equipment of spy literature: secret inks, peepholes, and periscopes. Surveillance photography (often taken by a buttonhole camera) plays an important part in establishing character and moving the plot along, and it is a photograph of the secret treaty, not the original, that is being smuggled out of the country. The film has all the fights and car chases the audience expected, as well as an ending in which the girl is saved, the villain thwarted, and his lair destroyed in one great explosion and fire—a narrative climax that the Bond films would always follow. *Spione* reflected real events, for Lang said the inspiration for the plot came from newspaper articles about Scotland Yard's raid in 1926 on a Soviet spy ring operating under the cover of a trade delegation in London.[2]

The emergence of a new Russian government intent on exporting the Bolshevik Revolution gave Cumming's service plenty of business in the 1920s, and he called it "capital sport." MO5 focused its attention on despicable Russian agents working underground in the United Kingdom, while MI6 sent brave patriots to spy on Russia and sabotage the communist war effort against the monarchists. Ian Fleming played a small part in this effort. Working as a journalist and stock broker justified frequent foreign travel, which included several trips to Moscow and Berlin, and this gave Fleming the opportunity to market his experiences to the press and the intelligence community. British intelligence regularly employed businessmen and journalists to gather information for them, and it is more than likely that his trips to Moscow were reported back to MI6. In this way he was more an asset or informer of the Secret Intelligence Service than an agent, which one KGB officer accused him of being in the 1960s. His colleague Sefton Delmer wrote: "I knew he was on some intelligence job or other. He was so much the type they used in those days. . . . It was clear he was no ordinary journalist."[3] This work provided Fleming with some experiences that later went into his Bond books, including the Russians' practice of placing hidden microphones in hotel rooms.

Dashing British spies outwitting the evil Bolshevik secret police in the snowy streets of Russia or thwarting the Soviet threat on the northern frontier of India provided some excellent material for stirring adventures, and these stories became very popular in the 1920s and 1930s. Robert Bruce Lockhart's *Memoirs of a British Agent* (1932) became an international best seller, and the semiautobiographical fictions of Somerset Maugham, Compton Mackenzie, and Paul Dukes gave the spy thriller a basis in fact. The title of the latter's book gives the general idea: *The Story of ST 25: Adventure and Romance in the Secret Intelligence Service in Red Russia* (1938). But it was the adventures of one agent in particular that cemented the idea of a master spy in the public's imagination.

Sidney Reilly had many of the trappings of an English gentleman, staying at the best hotels, eating at the best restaurants, and buying his clothes at the best tailors, but his charming manners could not fool Cumming, who referred to him as a "Jewish-Jap type . . . very clever . . . very doubtful—has been everywhere and done everything."[4] Cumming thought that sending him into Russia was "a great gamble," and Reilly's conflicts with the Bolsheviks eventually ended with his execution, but his fictional exploits made him into the first legendary figure of the British secret service and the first "real-life James Bond." One of his many wives promoted his image as the "Ace of Spies," and this was supported by the literary efforts of friends like Robert Bruce Lockhart, who called him "the mystery man of the British secret service . . . known today to the outside world as the master spy of Britain." Reilly was a modern spy. Pepita Reilly wrote that he entered German lines, "usually by aeroplane," to find out about secret military equipment, and one of his greatest coups was stealing plans of new German naval weapons in 1909. When Pepita's book was serialized in the *London Evening Standard,* her construct of the "Master Spy" entered popular culture and had some influence on Ian Fleming, who later commented: "James Bond is just a piece of nonsense I dreamed up. He's not Sidney Reilly, you know."[5]

The Bulldog Drummond stories of H. C. McNeile (Sapper) are another of the foundation stones of James Bond, who Fleming once said was "Sapper from the waist up and Mickey Spillane below" (referring to the American author of pulpy, violent, and sexually explicit detective stories). Fleming was introduced to the Drummond character as a schoolboy when the wife of the headmaster of the public school he attended read the stories aloud.[6] Drummond is a wealthy gentleman, formerly a British officer, who seeks out adventure in detective stories that owe something to Sherlock Holmes and Richard Hannay. A man of imposing physique, a crack shot, and trained in oriental methods of self-defense, Drummond had honed his skills in the Great War, where he went on solitary raids against the Germans. He is courageous, ingenious, and excels at sports and cards. Although he might resort to subterfuge, he is a gentleman and an honorable servant of the British Empire. Like Richard Hannay, he often finds himself accidently embroiled in an international spy ring from which he has to extricate himself and bring the perpetrators to justice.

Adventure stories and memoirs about spying were popular enough to attract the attention of film producers. Lockhart's *Memoirs of a British Agent* was made into a film, *British Agent,* by Warner Bros. in 1934, and several filmmakers, including Alfred Hitchcock, considered making films of Bulldog Drummond's adventures. Hitchcock developed a script called "Bulldog Drummond's Baby" in 1933, which was not produced because he could not obtain the rights to the character, but shortly afterward he was approached by the producer Michael Balcon, and their subsequent collaboration on five spy films made Hitchcock a household name in the 1930s. Hitchcock tapped the best spy fictions of the times—the work of Joseph Conrad, Somerset Maugham, Sapper McNeile, and John Buchan—and turned them into taut and suspenseful films, making his vision of secret agents and their spy craft the dominant one in 1930s and 1940s popular culture. "Bulldog Drummond's Baby" became the basis for *The Man Who Knew Too Much* (1934), although the Drummond-like hero was greatly watered down and lost much of the initiative and aggression that characterized McNeile's hero. The next film in Hitchcock's spy series, *The 39 Steps* (1935), was based on Buchan's 1915 novel *The Thirty-Nine Steps,* and this film defined the gentleman spy first imagined by writers like McNeile and Maugham: a handsome amateur who gets drawn into a dastardly plot but with a combination of force and ingenuity foils it. Robert Donat was the embodiment of the perfectly dressed, devilishly handsome, wisecracking agent/detective who takes on the villain single-handedly, saves the nation, and gets the girl in the final reel.

Hitchcock's film kept the core of Buchan's novel intact, especially the chase across the highlands of Scotland, but the "MacGuffin" driving the plot had changed significantly in the 20 years separating book and film.[7] In the book "the 39 steps" was a clue to the escape route of German spies, but in the film it is the name of their organization. In 1915, when Buchan published the book, the great secret, "a secret which meant life or death for us,"[8] was the disposition of the Royal Navy's Grand Fleet in Scotland. But in 1935 it was "a secret vital to your air defence . . . a new thing that many people are interested in . . . a new engine," which was so quiet that it made aircraft difficult to detect. The film accentuates the scientific nature of this secret by having a character called Mr. Memory recite the formula of the engine at the climax. Mr. Memory is a vaudeville performer who can answer any question put to him by the audience, and the spies use his encyclopedic memory to recall a formula full of unintelligible facts and numbers about rates of compression, dimensions, and angles.

After the triumph of *The 39 Steps* Hitchcock went from Buchan to Maugham and turned *Ashenden; or, The British Agent* into *Secret Agent* (1936). The hero was played by the Shakespearean actor John Gielgud (Donat was unavailable) in a plot that involved the assassination of a foreign agent. Gielgud was particularly flat in this unsuccessful film, but Hitchcock was drawn to the leading lady, Madeline Carroll. Hitchcock had already cast Carroll in *The 39 Steps* and had built up her role as the female character who accompanies and supports the hero, as well as falling in love with him—an early hint of the "Bond girl." But the coolly beautiful Carroll was no supporting character, and her self-confidence and wit made her at once the equal of the male hero and sometimes his better. Departing from Maugham's book, Hitchcock made Carroll's character a secret agent too, and she is much more decisive and adventurous than the understated Gielgud. When Carroll is asked in *Secret Agent* why she joined the British secret service, she responds: "Thrills, excitement, big risks, danger."

Hitchcock's contrariness and devilish sense of humor ensured that he would not stay within the formalities of spy fiction. In the last of his prewar spy films,

The Lady Vanishes (1938), the spy is female—an elderly spinster played by Margaret Lockwood. He also created a new kind of villain—not the monstrous Other who threatens violence and destruction but a polite, soft-spoken gentleman. The smooth-talking Professor Jordan of *The 39 Steps* is a bastion of respectable society. He was played by Godfrey Teale, who was chosen by Hitchcock for the role because of his uncanny resemblance to President Franklin Roosevelt. In *Secret Agent* the handsome, charming American (Robert Young) turns out to be a German agent, whose conversation is full of wisecracking Americanisms at first but then suddenly turns to barking orders in German. What appealed to Hitchcock was the duplicity, the turbulent evil hidden underneath the placid surface. Jordan tells Hannay: "I'm a respectable citizen. What would happen if it became known that I'm not what I seem?"

The Gentleman Spy

By the end of the 1930s British films and novels had established the conventions of the gentleman spy. Most often an amateur, the gentleman spy rarely exerts himself in his adventures and manages to fit feats of derring-do into a life full of travel, entertainment, fine dining, and beautiful women. Normally not an official of the established order, the gentleman spy often has a military background and enjoys the confidence and cooperation of men in high places. He sometimes has a father-son relationship with an authority figure who outlines the threat and directs the mission, but the gentleman spy works alone and only calls in the authorities once the crisis is over. His patriotism is self-evident as he works selflessly (and with no visible compensation) for the national interest against criminals, delinquents, and the machinations of foreign governments. Played by matinee idols like Leslie Howard and Robert Donat, he is dapper, urbane, and the perfect English gentleman. He is always impeccably dressed. Major Hammond, the gentleman spy in *Clouds over Europe* (1939), has a cupboard full of trilby hats and umbrellas to choose from before he leaves his luxury apartment.

Handsome and well-mannered, gentleman spies attract the attention of women but rarely chase them. They normally have steady girlfriends whom they treat with kindness and consideration. They do not play the field, and they prefer romance to seduction. They are members of the upper classes, independently wealthy (Bulldog Drummond is always accompanied by his manservant), and comfortable with the luxuries of the rich. They are refined on the surface but can defend themselves and their female companions when the need arises, but it cannot be said that they are violent men. They know how to use firearms but do not carry them around. When played by actors like Ralph Richardson and John Gielgud, the gentleman spy displays a typical British sangfroid and remains somewhat distant from the violence that is part of the job.

While the character of the gentleman spy hardly changed during the 1930s, the threats he faced became more militarized. The freelance secret agent took on crime in many forms, but as political events in Europe brought war closer, his enemies tended to be foreigners, often with central European accents, and the plots became focused on advanced military technology. Bulldog Drummond has to stop secret plans, new high explosives, and advanced bombers from falling into the wrong hands, which are often part of multinational organizations trading in advanced weaponry. In *Bulldog Drummond at Bay* (1937) the MacGuffin is a new type of airplane, a robot plane that is controlled by its inventor. The villains are in the pay of Ivan Kalinsky, "the mystery man of Europe," an arms dealer who runs his organization with modern tools like Dictaphones. His lair is a laboratory, full of glass retorts and test tubes, complete with a dentist's chair used for interrogation and torture. This room also serves as a gas chamber. A deluded scientist works the lever that turns it on in an alarmingly prescient scene that symbolically anticipates the role of German science and engineering in the Holocaust.

After the failure of the Munich Agreement of 1938 brought a war with Nazi Germany even closer, and anxieties about air raids were fanned by civil defense precautions that brought the whole population into the terror narrative, the plots of adventure films became more focused on the threat from above. It is invariably an advanced airplane, "American built . . . one of the new ones, capable of operating at high altitudes," that presents the threat. The plot of *Clouds over Europe* (released as *Q Planes* in Europe) involves enemy agents hijacking a top-secret airplane and its "valuable experimental apparatus." Other MacGuffins such as communication equipment and explosives are often made part of this threat from the air. In *Bulldog Drummond's Revenge* (1937) a new explosive called Hexonite has been stolen and must be retrieved. We learn that this explosive has been designed especially for aircraft bombs: a single airplane with 100 small Hexonite bombs has the potential to "wipe London off the map."

As the war drew nearer, the vocabulary of the films' plots became better informed and more specific, using correct technical terms like *supercharger* rather than making up some exotic name for the device that enables an airplane to fly faster and higher. The films express technological enthusiasm in the capital they confer on inventors and inventions. The invention is always foolproof, fully developed, and vastly more effective than the technology it replaces. The good scientist is invariably English and upper-class, and he (invention is exclusively a male occupation) often has to be rescued by the hero and his work protected. The bad scientist is always a foreigner and rather naive about the intentions of the organization for which he works.

As gentleman spies faced more ominous threats from new military technology, the stakes of the game rose accordingly. In the old days it was a matter of returning a valuable object or saving a single individual, but as dystopian science fiction and the strategic bomber merged into one reality in the late 1930s, "the safety of the country" often depended on the success of the mission. Secret weapons of unlimited destructive power brought together two heroic tropes in popular culture: the gentleman spies of spy films and the technological virtuosi of science fiction films. This is the foundation of the James Bond character.

Clouds over Europe anticipates the Bond films in many ways. The bad guys use a ray to disable the craft, force it down to the sea, and winch it aboard a disguised salvage vessel—a plot similar to that of *Thunderball* (1965). The film comes with spyware such as messages hidden in cigarettes, bad guys all dressed alike, and a major fight scene with scores of extras, which brings the film to its climax. It mixes together film footage of modern technology with special effects of action scenes using model airplanes and ships. It has two leading men (Ralph Richardson and Laurence Olivier): one is a gentleman spy who is a master of witty repartee, and the other is a handsome test pilot with a gorgeous girlfriend. One is an agent of the British secret service, the other a technologically literate man of action. It took a world war to blend them into one character and give him something really important to fight for.

War Films

War with Germany affected the stock characters of gentleman spy and evil scientist. Any doubt about the national origin of the scientist/villain was settled: he had to be German. The gentleman spy was now in uniform, but when he went on missions in enemy territory, he reverted to the traditional garb of trench coat and trilby. The secret weapons that form the MacGuffins of the plots were described more specifically as an alarmed public became more conversant with the vocabulary of terror. Most important, the war raised the stakes of the game, for the plots of world domination imagined in the sci-fi serials now looked like a reality in the ambitions of Adolf Hitler. With the world now on the brink of "a new dark ages," the secret agent had to save thousands and perhaps a whole city or country. In wartime the agent must risk everything because "it's my life against the lives of thousands." In the case of *Secret Agent* the thousands to be saved are the British forces in Palestine (a nice John Buchan touch), and the ultimate prize is victory.

Film studios in all the combatant countries made films that glorified their armed forces, but the emphasis on individual valor on the battlefield, which harked back to the films of the Great War, was tempered by the advance of military technology—the hero was now more likely to be a pilot or a submariner than an infantryman on the battlefield. Films like *Target for Tonight* (1941) and *In Which We Serve* (1942) showed fighting men operating complex weapons, and these heroes needed some technical aptitude to win the day. Although still leading from the front, the World War II hero was often encased in a machine as part of a technological system rather than a brave man going over the top.

The secret war also had a higher profile in the 1940s as resistance leaders and intelligence operatives were added to the pantheon of war heroes. *Desperate Journey,* made by Raul Walsh in 1942, has a group of downed American airmen led by Ronald Reagan carrying out a string of sabotage raids across Germany, destroying underground installations, chemical plants, and the factory making a new secret advanced airplane, "the Me114," which will give the Luftwaffe air superiority. They fight like commandos, creeping into enemy installations, disarming the guards, and sabotaging facilities with explosives. Once cornered, their escape is a Bondian clambering across roofs and dropping obstacles to confound following vehicles during a car chase, which is described as an "old Yankee bootlegger trick" and was destined for a long life in the Bond films.

The old clichés of spy films changed during wartime. The romantic train journeys, hidden identities, and romantic interludes seemed rather insignificant during a total war. Shady characters from international spy rings who lurk in dark alleys and secret rooms failed to reflect the reality of espionage in wartime. New intelligence organizations, such as SOE (Special Operations Executive) in England and the OSS (Office of Strategic Services) from the United States, produced new types of heroes and a new menu of action scenes to depict. The growing profile of resistance groups and special operations behind enemy lines redefined the war hero as a civilian operative of a resistance movement. Nowadays they might be called insurgents, but in World War II they were resistance heroes (and martyrs) who were honored in the wave of patriotic literature published after the war. This character represented the civilians pressed into service in times of national crisis—the inspired amateur was placed behind enemy lines in situations of great danger—and it gave a rare opportunity to make the war hero a woman. The life and heroic death of Violette Szabo, who was recruited into SOE, captured in France, and killed in the Ravensbrück concentration camp, was described in several books and a popular film, *Carve Her Name with Pride* (1958). A recent book about one administrator in SOE's French section carried Fleming's imprimatur: "In the real world of spies, Vera Atkins was the boss."[9] There were courageous intelligence professionals like Wing Commander F. F. E. Yeo-Thomas of SOE, whose exploits as "The White Rabbit" were made famous by Bruce Marshall's paperback book of the same name, which sold a million copies in the 1950s. British and American film studios produced a steady stream of films about this secret war: Lang's *Cloak and Dagger* (1946, starring Gary Cooper) for Warner Bros., Paramount's *O.S.S.* (1946, starring Alan Ladd), and 20th Century–Fox's *13 Rue Madeleine* (1946, with James Cagney). *Hangmen Also Die* (1943), *Manhunt* (1941), *Sabotage Agent* (1943), and *Secret Mission* (1942) all glorified the work of SOE.

Several of these American films employed the talents of expatriate German directors and writers, including Fritz Lang, Douglas Sirk, and Berthold Brecht, so it was not surprising that their work demonized the Nazis and created one of the most durable villains of film history—the arrogant Nazi officer who tortures, orders executions, and takes innocents as hostages. Real-life villains, like the SS leader Reinhard Heydrich, the "Butcher of Prague," were dramatized in war films. In *Hitler's Madman* (1943) and *Hangmen Also Die,* Heydrich threatens and postures, massacring hostages and sending innocents to concentration camps and young girls to the Wehrmacht's brothels. Nazi villains are usually masters of technology, from the secret laboratories of their perverted scientists to the radios and location devices used by the Gestapo to hunt down resistance fighters. In control rooms reminiscent of the sets of the 1930s sci-fi serials, they listen in to telephone calls and the bugs they have placed in rooms; they employ sophisticated listening devices to detect enemy aircraft; and they use electronic location equipment to search out spies. The resistance fighters who face them are few and not as well armed. They have to improvise equipment and usually win the day with a combination of ingenuity and heroism. They represent brave freedom fighters who take on the far superior forces and high-tech weapons of an evil empire, and it is surely no coincidence that the imperial forces in the *Star Wars* films often look and act like Nazis. Luckily for civilization and the fate of the brave resistance movement, for every "Death Star" the villains devise, the good guys find a way to overcome their technological disadvantage with technology of their own.

Fears of German science and their secret weapons continued through the war and into the immediate aftermath of the Nazi's defeat. *Cloak and Dagger* was based on a book that told "the inside story of General Bill Donovan's O.S.S." The protagonist is a physicist, Professor Alvah Jesper (Gary Cooper), who travels to find out how far the Germans have progressed in producing an atomic bomb. In Lang's screenplay the Nazis have successfully developed an atomic bomb, and Lang wanted the hero to find the secret laboratories deserted but with proof that the work had been transferred elsewhere—the point was that the threat of Nazi superweapons was not over. The studio changed the ending because it thought that the subject of atomic weapons was getting stale in postwar popular culture.[10]

German scientists had indeed been working on an atomic bomb, but this effort was sabotaged by members of the Norwegian resistance and British special forces, who blew up their supply of a vital ingredient called Heavy Water in one of the most famous commando raids of the war. The "Heroes of Telemark" were glorified in books and films, adding more luster to the exploits of irregular soldiers. The gentleman spy was being hardened by war and turned into a commando-like man of action. His primary job of gathering intelligence was supplemented by sabotage or even assassination. He began to demonstrate the range of skills expected of a soldier as well as a spy. He was a member of an elite group trained in special operations, which demanded skills in parachuting, swimming underwater, disembarking from small craft, and setting fuses for explosives. He was trained to resist interrogation and torture and not to make simple but fatal mistakes like carrying English cigarettes into German-occupied territory. He had to fire a sniper's rifle with telescopic sights or kill a sentry with a blow or a knife. The commando/spy acquired a much wider and more interesting range of equipment than the usual heroes of war films. He used ingenious devices such as a half potato to take duplicates of a rubber stamp, or a lighter combined with a cigarette case to make a camera. He was not a master of disguise like the 1930s gentleman spy, and he was much less urbane and much more violent than his predecessors. He lived a dangerous life and was a breed apart from the regular soldier.

These war heroes were celebrated in postwar epics about commandos and parachutists. One of them was *Red Beret* (1954), which follows the exploits of a group of parachutists and their raid on a secret German installation that followed the events of a wartime commando raid on a German radar complex at Bruneval quite closely. Made for $700,000, *Red Beret* returned $8 million in worldwide release and encouraged the producer, Cubby Broccoli, to follow up with *The Cockleshell Heroes* (1954), which told the story of a group of commandos who used collapsible canoes to penetrate German harbor defenses and sink ships. The plot was based on Operation Frankton, a 1942 commando raid on Bordeaux's harbor, and the film's long sequence of kayaking up the river anticipated many of the scenes in Broccoli's James Bond films when 007 must often undertake an arduous, dangerous trip into enemy waters. Both films act as vehicles to explain the role of irregular forces, describe their training and equipment, and glorify their individual exploits. *The Cockleshell Heroes* had all the heroics expected from a suicide mission, and Simon Winder thinks it succeeded because "it clearly mapped British self-image perfectly," with its themes of amateurism, voluntarism, self-sacrifice, and the pitting of an individual against a powerful foe. Winder sees this film as the beginning of Broccoli's successful reading of the national mood, which would continue into the 1960s with his phenomenally popular James Bond films, and as a summation of everything that Fleming had done in the war "and everything he wanted from Bond."[11]

SOE organization was typical of a nation that is not military minded, does not win first battles and learns by trial and error. Schoolmasters and accountants became saboteurs and instructors of secret armies.

Edward Cookridge, *Inside SOE*

006 Ian Fleming, Intelligence Officer

Fleming was one of the hundreds of businessmen, journalists, academics, and playboys recruited into the intelligence services at the beginning of World War II. This was one of the great mobilizations of the war, and it opened up the world of secret intelligence to an army of enthusiastic amateurs like Fleming. As assistant to the director of Naval Intelligence his job was to liaise with the other military intelligence branches and also with MI5, MI6, and the government code breakers. This widened his network of friends and informants and brought him a bird's-eye view of intelligence operations, which he was able to exploit later in his James Bond books. His Section 17 was the coordinating section, the nerve center of Naval Intelligence, and the ideal place to exercise his charm and social connections in service of the war effort and a little bit of excitement. Rear Admiral John Godfrey was impressed with the young stockbroker's ability to digest a wide range of information and "arranged for him to be shown everything."[1] Fleming also carried out intelligence planning, which gave him free reign to combine technological enthusiasm with his vivid imagination to come up with a stream of clever (but often impractical) ideas to help win the secret war. One of his ideas involved the infiltration of a group of islands off northern Germany, "whereby I and an equally intrepid wireless operator would be transported to the group by submarine and there dig ourselves in, to report on the sailings of U-boats." He also planned to highjack one of the fast motorboats the Germans used to rescue their pilots downed at sea. These "Red Indian day dreams" as he called them later were remembered fondly as he sat by his writing desk in Jamaica and thought up James Bond's next piece of derring-do.[2]

Fleming's work for Naval Intelligence brought him into contact with the commandos and irregular soldiers who carried out special operations for the secret service. Fleming's biographer and colleague at the *Times*, John Pearson, remembered: "London was full of secret organizations. Private armies were forming everywhere. The geniuses and the crackpots were getting their chance at last." Pearson thought that it was "the fantasist in Fleming" that drew him to these private armies.[3] Both Fleming and Churchill were men of action who admired audacity over stealth, brawn over brain, and nowhere was this clearer than in Churchill's beloved Special Operations Executive. SOE was formed in the summer of 1940 during the low ebb of the war for the British Isles; it was a strategy created out of defeat. With France overrun, and the British Expeditionary Force (BEF) regrouping to defend the homeland, the government grasped at any opportunity to take the war to occupied Europe. The only options available were the bombing raids carried out by the RAF and irregular warfare, which could nibble at the edges of the great German Reich. Famously tasked by Churchill "to set Europe ablaze," SOE was to sabotage and harry the Germans in occupied Europe and build up the nascent resistance movements there, a brief that was similar to T. E. Lawrence's duties in Palestine and was certainly influenced by his writings about indigenous resistance movements. SOE operations were directed by Lt. Col. Colin Gubbins of the Royal Artillery, who was given the identification symbol *M*, later to be appropriated by Fleming for the director of MI6 (who had always been known as "C" in the real world, from the days of Mansfield Cumming). The *Partisan Leader's Handbook*, written by Gubbins, reflects Lawrence's thinking and the tactics of the Boers: "Surprise is the most important thing in everything you undertake. . . . Mobility is of great importance. Act therefore where your knowledge of the country . . . give[s] you an advantage over the enemy. Never get involved in a pitched battle."[4]

These private armies needed special equipment. The Royal Navy was responsible for transporting SOE agents into enemy territory, and as part of his duties Fleming obtained the submarine HMS *Truant* to insert a commando "striking force" (which included his brother, Peter) into Norway. The navy was asked to provide transportation for so many of these missions that several submarines were converted to specialize in covert landings. The Bond films carried on this tradition by often using a Royal Navy submarine to carry Commander Bond on his secret mission. The navy used motorboats for offensive raids on the enemy's coast. These high-powered craft could make up to 28 knots and were the sort of machine that attracted technological enthusiasts like Fleming—streamlined, powerful, and exhilarating to pilot as they went crashing through the waves. His detailed descriptions of fast boats appear regularly in the Bond books.

The war brought Fleming into contact with a plethora of fascinating gadgets: miniature submarines, sniper rifles, coding machines, and aerial cameras.

Cameras for aerial photography had improved markedly in the 1920s and 1930s, especially thanks to the work of Sherman M. Fairchild in the United States, who designed a between-the-lens shutter that produced much clearer images. The Fairchild Company went on to become an innovator in this field, but on the other side of the Atlantic there was little done to advance this technology. The RAF showed little enthusiasm for aerial surveillance, so in 1938 Frederick Winterbotham recruited Sidney Cotton to fly over Italy and Germany in a private Lockheed 12A aircraft that had been acquired with SIS funds. Cotton and Winterbotham had to make alterations to this airplane to hide the cameras and devise special mounting frames and automatic control equipment. They experimented with combining several cameras in an oblique format, which enabled wide areas to be photographed. In a story of brilliant ingenuity that often went against conventional wisdom, these two amateurs came up with photographs so impressive that potential customers found it hard to believe them, but soon "the excitement was intense. Requests came pouring in from the Admiralty, the War Office and the Air Ministry."[5]

In the first weeks of the war one of Fleming's colleagues in the Admiralty told Sidney Cotton about the need to get a photograph of the German navy at anchor in Wilhelmshaven. Cotton obliged, and it was Fleming who brought these photographs to the office of the First Sea Lord, raising Fleming's status within the Admiralty and cementing a fruitful friendship between the two men. Fleming's biographer, Andrew Lycett, writes that "Cotton became the first in a line of practical inventors befriended by Ian." They met for brainstorming sessions, and as Cotton explained his latest gadget, Fleming would respond with "Amazing, Sidney, amazing."[6] Cotton was nominally an employee of Winterbotham's Air Intelligence branch within SIS, and the latter was protective of his personnel and wary of Fleming's attempts to recruit Cotton. Despite the wartime footing, interservice rivalry was still strong, and Winterbotham insisted that aerial photography remain exclusively within the domain of the RAF. Aerial recon photos, called "covers," became one of the most valuable currencies of the secret intelligence war and were jealously guarded to the point of being deliberately hidden from other departments, which might have made good use of them.

Cotton was essentially a maverick who liked to work outside the rules of officialdom, and this suited Fleming, too. They conspired to embarrass Winterbotham with an unscheduled flight over Portsmouth, which proved that the Air Ministry's (preradar) aircraft-warning system was inadequate. His independent outlook, irregular work habits, and ability to get things done outside the bounds of bureaucracy offended the Air Ministry establishment, but it brought him the support of Air Chief Marshall Sir Cyril Newall, who told Cotton, "Your value to me lies in your continuing to do things in your unorthodox way. I want results."[7] Cotton was described by R. V. Jones as "a buccaneering Queenslander," and he was the kind of inventor that Churchill and Fleming admired: independent, resourceful, and, most important, a practical man who often came up with simple but effective solutions to problems. One such problem was the condensation that built up on the lens of the cameras as they operated at higher and colder altitudes—and ruined the photographs. When Winterbotham and Cotton first tested their cameras in the Lockheed, they were amazed to get "absolutely clear photographs taken from over 20,000 feet." Winterbotham found the answer when he checked the small flap that covered the camera hole; warm air from the heated cabin was flowing under the airplane and over camera lenses: "Here was the explanation for the absence of condensation, and the secret of our success. It was as simple as that. . . . Both Cotton and I realized that this was something of supreme importance . . . for the whole future of aeronautical photography." From then onward Cotton routed exhaust gases from the engine over the camera lenses.[8] Cotton also solved another problem for the pilots taking pictures with obliquely angled cameras; it was difficult to see the target because pilots had to twist around in the cramped cockpit to look over the aircraft's side. Cotton invented a new "tear drop" molded Perspex window with a bulge large enough to enable the pilot to look out of the side of the cockpit canopy. This simple innovation was immediately adopted, and the Triplex Company made one hundred thousand of them during the War.[9]

Sidney Cotton made too many powerful enemies to survive long in the bureaucratic infighting of the war effort, and his civilian photoreconnaissance operation was soon taken over by the Air Ministry. The RAF split the job into two parts: one unit (PRU) took the photographs, and another (PIU) interpreted the pictures. Little effort had been made by the RAF's photographic officers to develop the skills of interpretation, which was done by staff with equipment no better than magnifying lenses bought from local jeweler's shops. The Central Interpretation Unit at Medmenham was charged with finding, collating, and interpreting the photographs taken by the PRU's Supermarine Spitfires and De Havilland Mosquitos—about a million a month at the height of the conflict. They dealt with hundreds of requests for images each day and ended the war with a staff of more than 3,500. The staffing and organization of Medmenham exhibited the same "typically British, amateurish muddle and improvisation" that characterized the mobilization of scientific and creative resources during the war. The archaeologist Glyn Daniel was recruited into PIU and found "an ill-assorted collection of dons, artists, ballet designers, newspaper editors, dilettanti, writers . . . the splendid inspired madhouse of civilians and RAF and WAAF officers interpreting air photographs, making photographic mosaics, sending out annotated photographs."[10]

The staff at Medmenham were to discover some of the most important images of the secret war and played a critical role in providing intelligence about Germany's radar technology and missile development. Aerial photography emerged from the war as a major part of intelligence gathering, but Fleming rarely mentions it in his Bond novels, and it never provides the quality information to Bond that it did during the war. M has the Canadian "frontier patrol planes" take "a full aerial survey" of the villains' lair in *For Your Eyes Only,* but it comes up with nothing useful. Perhaps the interservice demarcation battle that he lost to the Air Ministry put him off including it in his spy stories, and Fleming himself admitted, "I hate photographs and taking them."[11] In the books, Bond usually manages to spot something that all aerial reconnaissance misses, but in the real world such reconnaissance was going to transform the gathering of intelligence and eventually make Bondian secret agents obsolete. Aerial photography and code breaking were proving their worth and overshadowing the contributions of individuals working in the field. World War II proved to be the watershed in machine-based intelligence, and from then onward code breaking and aerial surveillance were to account for the bulk of the budget for intelligence operations. Fleming's technological enthusiasm, at least in his writing, was always tempered by a nostalgic reverence for the past—its values and its heroes. However much Bond may be a modern man, with every technological tool at his command, the character harks back to the romantic adventures of Sidney Reilly and Richard Hannay. Fleming would much rather tell a good story based on ingenuity and violent action than give prominence to the scientific work that goes on in the backroom of an office building. Although Fleming was party to the "Ultra Secret," code breaking appears only once in Bond's adventures, as the MacGuffin in *From Russia with Love.*

Despite the well-publicized triumphs of breaking the enemy's codes, British intelligence still had one foot in the nineteenth century with some ancient equipment, including carrier pigeons! It also employed some ancient espionage tricks. Operation Mincemeat, described in a popular novel and film as "The Man Who Never Was," misled the Germans about the Allied invasion of Sicily. A dead man was dressed up in the uniform of a naval officer, complete with fake plans for an invasion, and dropped off the Spanish coast by submarine, where the body was discovered and handed over to German intelligence. This updated version of the "haversack trick" was devised by Fleming's colleague in Naval Intelligence, Ewen Montagu—a man and a story after his own heart.

At war's end a case could be made for both technological and individual contributions to the victory of the intelligence war. The D-Day landings were the largest amphibious operation ever mounted, and they required a massive intelligence effort that encompassed thousands of decrypted signals and millions of photographs. All units tasked with landing on a beach had a scale model of their target (created from photographs) to examine before they attacked, and it is said that soldiers stopped in the middle of combat on the beaches to examine their aerial photographs.[12] The massive deception plan that misled the German High Command into thinking that the Allied invasion of France was to be at Pas de Calais, rather than the beaches of Normandy, involved many aspects of the secret war, from the electronic networks that built up the radio identities of armies who were not there to actors who impersonated English generals who were really somewhere else. It covered the work of the engineers who built dummy tanks and aircraft and the feathered friends of MI5's Pigeon Service Special Section, who carried fake questionnaires about German defenses to the Pas de Calais to be picked up by German soldiers.[13] Yet the greatest single contribution came from a double agent named "Garbo," who convinced German intelligence that the landing on the Normandy beaches was just a feint and that the real attack (from the imaginary armies described above) was to land at Pas de Calais in the next few days. This held up reinforcements of the troops at the beachhead and brought confusion and delay to the Nazi response to the invasion. Garbo was a Spanish civilian called Juan Pujol, who rightly claimed that "the pen is mightier than the sword." He fought a war of words—a total of 315 letters in secret ink, and 1,200 wireless messages over a period of three years.[14] His words changed the course of the war and were a reminder to Fleming and his colleagues that the ingenuity of human agents still had a role to play in the secret war.

The Gadget Factories

Fleming's work in Section 17 of Naval Intelligence also earned him access to secret weapons development, which was going on in university laboratories, government workshops, and the backyards and kitchens of independent inventors all over the country. The British war effort marked an unprecedented mobilization of science and technology, which the prime minister embraced wholeheartedly. Churchill the technological enthusiast was an innovator whose support of radical new weapons like the "landship" had brought him into conflict with the military's innate conservatism. In the Great War the officer corps called his idea for an armored vehicle with caterpillar tracks "Churchill's Folly," but the tank turned out to be a success, and this gave Churchill an appreciation for independent ideas and inventors and the contributions of practical engineers like Eustace d'Eyncourt and Ernest Swinton, who made the "landship" work. As a lover of gadgets and practical mechanics, Prime Minister Churchill threw his support behind a wide range of projects during World War II, from using large bodies of ice as aircraft carriers to automated machines that could quickly dig out trenches. David Edgerton argues that "wartime Britain saw an extraordinary cult of invention and the inventor, whose high priest was the prime minister himself. Gadget factories of all sorts flourished under his leadership." A. V. Hill was thinking about Churchill when he commented in 1942: "There have been far too many ill-considered inventions, devices, and ideas put across, by persons in high places, against the best technical advice."[15]

Gadget factories sprang up like mushrooms all over the country. Military Intelligence ran a small research section, General Staff Research, into irregular warfare under the direction of Colonel J. C. F. Holland. This section GS was turned into MI(R) (Military Intelligence Research), and Holland put a subsection MI(R)c under the command of Millis Jefferis of the Royal Engineers and gave it the task of developing weapons for irregular warfare. At the same time, the "cloak and dagger boys" of SIS formed a Section D (for destruction) under Major Lawrence Grand to look into methods of sabotage and subversion. Monty Chidson was Grand's deputy, and he put together a group of saboteurs and adventurers who carried out some of the war's first coups de main: the rescue of Madame de Gaulle (the wife of the leader of the Free French) and the hurried evacuation of the diamond reserves at Amsterdam, both adventures in the Bondian style. With the help of Guy Burgess, who had been recruited from the BBC, Chidson set up a training facility for saboteurs.[16] MI(R) and Section D were amalgamated and brought into SOE, but MI(R)c remained independent, and as MD1 it continued to develop equipment for guerilla warfare. An officer in MI(R)c pointed out the difference between them: Jefferis and his uniformed officers were "a more or less legitimate outfit which was to produce unusual or respectable weapons . . . whilst Grand & Co. (Section D) would get double the money for running a Cloak and Dagger outfit." MD1 moved from the War Office and established itself at a country house called "The Firs," near Aylesbury in Buckinghamshire. It was nicknamed "Churchill's Toyshop" because of the prime minister's enthusiastic support: "In order to secure quick action, free from departmental processes, upon any bright idea or gadget, I decided to keep under my own hand . . . the experimental establishment formed by Major Jefferis . . .

this brilliant officer, whose ingenious inventive mind proved . . . fruitful during the whole war."[17]

SOE brought in scientists and engineers to help equip its agents, creating "a sequence of laboratories which soon supplied a fantastic variety of special equipment needed by the agents to survive—and in certain circumstances to die."[18] There were separate sections set up to deal with espionage requirements such as forgery, explosives, and communications. The establishments charged with the design and production of spyware were designated by roman numbers.

Station XIV specialized in forgery of identity papers, ration books, and all the documents needed to travel in occupied Europe. Staff from banks and the printing companies that produced banknotes, postage stamps, and shares was recruited to mass-produce forged documents at a specially equipped printing works. The venerable printing company Waterlow, supplier of banknotes to the Bank of England, lent its expertise to this effort. The assistance of Scotland Yard (HQ of the Metropolitan Police) was sought in recruiting the best forgers in the land, and in this way convicts were able to serve the war effort by producing large amounts of currency and official documents in return for reductions in their sentences. In the early days forged papers for agents were of poor quality and described as "death warrants," but by 1943 this section was producing documents that could stand up to scrutiny.

Station IX, at a former hotel called "The Frythe," carried out research on radios and took over the weapons research work of Station XII at Aston House in Hertfordshire, which Fleming is known to have visited early in the war. Station XII then became a manufacturing and storage center for weapons, explosive devices, and booby traps, while Station IX became the research and development arm. Professor Dudley Newitt was made director of its Scientific Research Department. Newitt was a chemist who served in Palestine in the First World War, a member of the Royal Society, a professor of the Imperial College of Science, and a leader in the field of high-pressure technology. He gathered around him scientists with equally impressive resumes: Dr. Douglas Everett had advanced degrees in chemistry from Oxford University, and the physicist Professor E. G. Cox came from the University of Birmingham and was a pioneer in mapping the crystal structure of explosives. By 1942 this Experimental Section employed more than a dozen scientists, some pure academicians and some industrial researchers. Dr. F. Freeth had been a director of Imperial Chemical Industries (ICI), Alkali Division, and had carried out research that contributed to the invention of polythene. Station IX also had the Engineering Section, under the command of John Dolphin, an academically trained engineer who had invented a coal-cutting machine while managing the Austin Hoy Company. He was assisted by John Meldrum, who had first-class honors science and engineering degrees, and by Hugh Reeves, a graduate from Cambridge University. In their book on SOE, Fredric Boyce and Douglas Everett noted that in addition to these engineers the Engineering Section employed "194 tradesmen," an indication of the separation of craft and academic skills in the scientific hierarchy of the 1930s and 1940s.[19]

The Engineering Section had several subsections, each devoted to weapons, explosives, incendiaries, fuses, transportation, and underwater equipment. There was also a Physiological Sub-Section headed by Dr. Paul Hass, a professor of plant chemistry at University College, London, and a lecturer at the Royal Botanical Gardens at Kew. He was later joined by Dr. Sandy Ogston, a biophysical chemist and a Fellow of Balliol College, Oxford. Ogston developed the compact ration packs issued to agents (which employed the latest techniques of dehydrating food) and medical kits that could fit into a flat cigarette tin. He worked with Geoffrey Bourne and Dr. Ken Callow, who both had distinguished postwar careers in the science of nutrition.

Station XV, at "The Thatched Barn," provided agents with the appropriate local clothing and personal effects—the pocket debris to support the cover story of the operative. Wartime rationing in occupied Europe meant that some items of clothing, such as fine English-made leather shoes, were rare and unusual on the continent, as were American cigarettes and matches. Before being sent into the field, each agent's clothing was literally cleaned with a toothbrush to ensure there was no trace of any incriminating material, such as Virginia tobacco, which might give the game away. One of the leading figures at Station XV was Elder Wills, who had been recruited from the British Expeditionary Force in France, where he had worked on camouflage. Wills and his staff, many drawn from the film industry, devised ways of concealment and hiding espionage equipment in everyday items. They provided custom-made camouflaged containers for explosive devices, which ranged from dead rats to bicycle pumps. One of their first tasks was to find old weathered suitcases to house the wireless sets that SOE operatives carried into occupied Europe. The general-issue suitcases they were given were all the same and proved to be an easy way of identifying SOE wireless operators. All the German military service (the Abwehr) had to do was to have someone at the railway station who could recognize the British suitcases and arrest anyone carrying them.

The Special Communications Section of Station XV concentrated on escape equipment. Major Clayton Hutton, a film director in civilian life, devised hiding places for tiny maps, compasses, and "escape saws" in buttons, signet rings, and pipes. This section also made the escape maps used by agents and by downed Allied airmen to find their way out of occupied territory. One member of this section recalled: "We used to do the little silk maps to go inside handbags . . . and to roll up and put inside fountain pens." The Thatched Barn had an art department and a "dirtying shop" to weather the items they produced. Special laboratories were custom-built for the handling of explosives: "No concrete dust. Linoleum, half inch cork liner, which we could not get in England. . . . Somehow we got it. So if anything dropped there was a softness."[20] One of the infamous products of this laboratory was the explosive turd, which was designed to disable German vehicles as they drove over it. Experts from the London Zoological Society and keepers of the Natural History museums were enlisted to make the devices as realistic as possible, and all manner of horse, camel, mule, and elephant droppings were examined. Another section shaped explosives that went into ingenious booby traps like explosive rats and pieces of coal, which were meant to be thrown into the coal supply of a locomotive, ship, or factory.

Station XVII was a school of industrial sabotage run by George Rheam, who has been called "the father of modern industrial sabotage."[21] He educated SOE personnel to locate the essential machines in any factory, destroy them with the minimum of explosive, and make sure that repairs could not be made promptly. Station XVII determined the size and type of the charges used in specific sabotage missions, such as the attack on the heavy-water plant in Norway. The ever ingenious Colin Gubbins recruited claim adjusters from the insurance companies to lecture the students at this school for sabotage; these men had real-life experience in the action of fires and explosions.

Science versus Technology

The advances in aerial photography, code breaking, and nuclear energy remained top secret in the years immediately after the war, while the exploits of the men in the gadget factories were well loved and well known. At the beginning of the war it was fashionable to ascribe a string of embarrassing defeats to the inferiority of Allied weapons, for as the deputy chief of the Imperial Staff conceded: "German equipment is as good or better than ours." Allied soldiers thought that every tank that opposed them was a Tiger and every machine gun they faced was an MG 42, a weapon with such a high rate of fire that the US Army produced a training film to counteract the fear it instilled. With victory in sight there grew a patriotic pride in the British ingenuity that had overcome a vastly superior enemy in terms of both quantity and quality of arms. The nation liked to think that the inspired amateurs and eccentric inventors had triumphed against the odds, if only just. Yet, as the secrets of World War II were gradually revealed, there was a realization that the British had managed to develop some advanced technology into war-winning weapons, such as radar and the atomic bomb. Science had proved vital in the defense of the United Kingdom against air attack and in the Battle of the Atlantic, so there were grounds to believe that science had won the war for the Allies.

Science also triumphed in the greatest intelligence coup of the war—the breaking of German codes. This revelation was not made available to the public until the 1970s, when Winterbotham was allowed to publish his account. As a member of MI6 (he reported directly to its head, Sir Hugh Sinclair, and then his successor in 1940, Sir Stewart Menzies), Winterbotham was fully involved in what was called "The Ultra Secret." The intelligence derived from Enigma decrypts was often of immense value, and the histories and films about the work of the code breakers of Bletchley Park have focused on brilliant young scientists like Alan Turing. These narratives overlook the engineering and craft skills of post-office engineers who built the machines that ran through the almost infinite number of permutations to come up with the keys of the Enigma codes. David Edgerton has argued that this emphasis on science has turned attention away from the major contribution that came from industry, business, and engineering. In his opinion World War II was "The War of Experts": "Britain was moderately good at inventing, but very good at developing."[22] This distinction between "science" and "mechanics," in the words of Winston Churchill, was followed by Fleming in his Bond stories, as he favored amateur inventors and practical engineers like Cotton and Jefferis over the scientific backroom boys. Fleming's inspiration came from the practical men who equipped irregular troops rather than from the theorists who represented a scientific elite. He leaned toward characters like Stuart Macrae and tended to go along with his jaundiced view of the "boffins" at Bletchley Park: "brilliant, long-haired youths who contrived to break the enemy's codes every so often."[23]

"If it hadn't been for Q branch, you'd have been dead long ago."

Q to Bond, *Licence to Kill*

007 Equipment

The character of Bond's quartermaster, Q, comes from Fleming's war experiences. The first model for Q was Charles Fraser-Smith, who worked for the Ministry of Supply, devising equipment for secret operations. He remembered that Fleming "was always keen on gadgets" and interested in his work. Fleming liked to visit the clandestine gadget factories and sometimes got hold of special equipment and tried them out on his colleagues in NID.[1] Major Macrae of MD1 remembers seeing him at the Firs workshop. John Pearson also noted Fleming's interest in the weaponry of the secret war—from the bicycle generator that powered wireless sets to the fountain pen that poured out tear gas when the clip was pressed. Fleming carried the pen with him when he traveled overseas.[2] These are the sources for the gadgets that Fleming added to his stories to make them appear more realistic, another way he turned his wartime experiences to advantage in his literary career.

Charles Fraser-Smith was recruited initially as a procurer of equipment and supplies; his job was to obtain materiel for agents going overseas. He had no previous experience in covert operations and came into the war as a civilian farmer. His job required a blend of mechanical ingenuity and ingenious scavenging. For example, Fraser-Smith acquired French tobacco to supply to his agents and used the extensive collection of foreign matchbox covers assembled by the British Bryant and May match manufacturer as examples for his forgers. He proudly recalls that even the matches taken into the field had been properly researched and sourced. Such attention to detail was vital because it took only a small slip to betray an agent to the watchful enemy. One of his first tasks was to secure a supply of the miniature cameras that had become one of the basic tools of espionage, replacing the large and noisy Photostat duplicating machines in general use. The best miniature camera was the Minox, which produced 50 well-defined images on a film about one-sixth the size of a postage stamp. Introduced in 1937 by the Valsta Company of Riga, Latvia, it was small enough to fit in the palm of one's hand and was easily activated by pulling the sliding case out. The Minox III was possibly the single most important piece of equipment in twentieth-century espionage. Fraser-Smith was charged with obtaining a large number of these cameras in a hurry, but he could not find enough to meet demand, so he approached Meccano Ltd., a well-known manufacturer of toys, whose small model "Dinky Cars" required precision machining of many tiny parts. Fraser-Smith contracted with Meccano to make a copy of a Minox. He also made an arrangement with Kodak to manufacture film for these cameras. Once the design was finalized and production facilities set up, these Minox copies were modified to fit into concealment devices. Fraser-Smith liked to use cigarette lighters for concealment because they could be operated in public without attracting undue attention.[3]

Photography was an important skill for a field operative, and it was taught at SIS training schools, along with "arrest information," first aid, security, and unarmed combat. Bond is a capable photographer, using a Leica with a flashlight to take pictures of documents in *Moonraker*, the same equipment employed by the Chinese agent to take his picture in *Doctor No*. By the time Bond came along, photographs were an essential part of the data collected by intelligence organizations. They were used in the German service depicted by Fritz Lang in *Spione* and have been instrumental in building up reference files in every intelligence service since the First World War. When Adolf Eichmann of the Gestapo's Jewish section visited the Middle East in 1937, he was photographed and the image placed in the SIS files.[4] Fleming knows something about cameras, for he mentions the "Speed Graphic" used by news photographers, but cameras do not play any significant part in Bond's operations, where they are mainly used to conceal other devices such as tape recorders or Geiger counters. The famous Minox "spy camera" does not receive a mention in the books but appears twice in the films. Bond has a Minox A/IIIs in *On Her Majesty's Secret Service*, and the Minox he uses in *Moonraker* has been customized with an engraved 007 with the middle *o* over the lens. By 1979 the brand was as important as the gadget.

Fleming worked with Fraser-Smith on his audacious plan to create a Q plane, a German bomber manned by RAF and SIS personnel that was intended as a trap for the Luftwaffe's air-sea rescue boats and their precious codebooks. Fraser-Smith was commissioned with supplying the appropriate German uniforms. Over time Fraser-Smith moved from obtaining supplies to designing concealment devices: compasses in pens, Minox film in shaving brushes, and hidden compartments in torches, tubes of toothpaste, pipes, and so on. He did much of his work for MI9, which was coordinating the important effort of rescuing downed aircrew from occupied territory and providing escape tools to prisoners of war. Fleming was intrigued with these ingenious devices, which took personal items such as shaving brushes or shoes and carved out secret cavities in which escape equipment like tiny saws or compasses could be concealed. Some of Fraser-Smith's ideas were translated into the gadgets that Q issues to Bond in the books and films. Fleming, an avid golfer, specifically asked Fraser-Smith about his work in concealing compasses in golf balls that were sent in Red Cross parcels to POWs. Fleming made use of the information in *Diamonds Are Forever,* where the smuggled gems are concealed in Bond's Dunlop 65s, but as Fraser-Smith points out, if "any suspicious investigator tried to bounce Fleming's balls they would have given their carrier away instantly." Fraser-Smith's golf balls had been redesigned with an extra rubbery interior to bounce just like the real thing.[5]

Q, like Bond, is a composite figure drawn from men Fleming had known or heard about. Concealment was Fraser-Smith's métier, but that only describes a part of the work done by Fleming's Q. In a postwar analysis of SIS operations, the deputy director of the service, Bill Cordeaux, commented that "ingenious concealing devices are a favorite subject with writers of spy stories. . . . In practice there is not a great deal to say on that subject. Once a person is suspected of carrying secret papers, it is almost certain that a rigorous search will discover them." Concealment devices were useful only for routine examination and therefore did not have to be too complicated or devious. One important point that Cordeaux made was that most documents were now "photographically reduced in size," which helped in their concealment. Microphotography was a boon to men like Fraser-Smith.[6] Fleming's Q provides us with few examples of concealment—the familiar attaché case with secret pockets will normally suffice in the books. But his main concern is with explosives and special weapons, and this takes us away from Fraser-Smith of the Ministry of Supply to the much more warlike workshops of SOE and MD1. The rise of irregular warfare had shifted the role of spyware from concealment to the technology of sabotage and assassination, and this is where James Bond comes in.

Irregular Research and Development

The rise of MD1 to become "Churchill's Toyshop" provides an example of the technical abilities admired by Churchill and Fleming and a contrast with the more scientific work of other experimental departments of the secret war. Jefferis and his staff were as irregular as the operatives they equipped, strong in ingenuity but somewhat lacking in academic credentials. Jefferis was called a genius by many who knew him, but he lacked administrative and diplomatic skills and behaved in a manner that his peers would have called "eccentric." His second in command, who ran the show and managed its exponential growth, came not from Oxford or Imperial College but from the editor's desk of a magazine called *Armchair Science.* Stuart Macrae had apprenticed at the Westinghouse Brake and Signal Company and got a job as a draftsman at a small aircraft manufacturing company. During the Great War his services had been employed by the Air Ministry's experimental station in the design of bomb racks for bombers. After the war Macrae had made a living as a technical journalist before he received a fateful phone call from Jefferis in June of 1939 inquiring about some new magnets he had read about in *Armchair Science.* Jefferis was interested in devising a magnetic mine that could be attached to ships, and it was this project that brought Macrae to MD1.

Macrae could not provide Jefferis with any of the new Alnico magnets being developed in the United States, but his assistant alerted him to some smaller versions he had seen in the window of an ironmongers' shop nearby, and Macrae immediately bought the entire stock. Next, Macrae needed a workshop, and this brought him to C. V. "Nobby" Clark, who had fought in the Great War and spent the peace running his Low Loader Trailer Company. While editing *Caravan & Trailer* magazine, Macrae had seen some of Clark's ingenious trailers and was impressed by "his ability to view mechanical problems in an unorthodox way." Macrae and Clark set themselves up in the latter's house and began work on the magnetic mine. They bought some large tin bowls from Woolworth's and paid a tinsmith to make some connecting rims to connect up the casing for the magnets and explosives. They then constructed a belt with a steel plate (the size of the ring of magnets on the casing), which attached it to the swimmer. Using porridge from Clark's kitchen to act as the explosive filling, they adjusted the weight and buoyancy of the mine until it worked well enough in a bathtub to be taken to the local public baths for field testing. The next problem, and the main technical challenge of designing any sort of mine, was the delayed-action firing device. Macrae and Clark came up with a spring-loaded striker held in place by a pellet soluble in water, which they hoped would give the swimmer a couple of hours of delay after attaching the mine to a ship. They searched in vain for a material for the pellet that would dissolve at a regular rate, until one of Nobby's children obliged by spilling some aniseed balls on the floor. Macrae discovered (by popping a few into his mouth) that these sweets always dissolve at the same rate. Macrae claims to be the first man to drill a hole in an aniseed ball, and after doing this, he attached it to the igniter and found that it worked perfectly. They then rushed into town and bought up entire stocks of aniseed balls in every sweetshop they could find. When they contracted with the sweet maker, Messrs. Barratt, to mass-produce aniseed balls for them, they found that Barratt made them so precisely that "nothing can be more alike than two aniseed balls." Finally, they needed a watertight cover to keep this device dry before use and discovered that putting a contraceptive over it was a quick and easy way to do the job. They made another trip to town and bought up all the stocks in chemists' shops "earning ourselves an undeserved reputation for being sexual athletes."[7]

Their invention was called the limpet mine, and it became one of the most common sabotage devices used in the war. The size of a tin helmet and "shaped like a bowler hat . . . and when the thing exploded it blew inwards," limpet mines were used extensively in the war and the uneasy peace that followed, blowing up ships in harbors (including those taking Jewish immigrants to Palestine) and any target made of metal. Bond uses limpet mines in *Thunderball* and *Live and Let Die,* and one nearly kills Kerim Bey when it is attached to the wall outside his office in *From Russia with Love.* Limpet mines were still in use in Bond films by the time of *Tomorrow Never Dies* (1997), when they are stocked in Wai Lin's bike shop. The kit that Bond carries in *Thunderball* when he sets off to reconnoiter Largo's boat is typical for a commando raid in the war: Fairbairn and Sykes dagger, limpet mine, "a dozen pencil-shaped metal and glass fuses," torches, and Benzedrine tablets.

The limpet was such a useful device that Macrae produced a much smaller version, only 6 × 4 inches, which was called a Clam. By the end of 1939, MD1 had eleven different devices (mines, grenades, switches, and booby traps) under production and was developing five more. As the workload increased, Macrae began to engage outside contractors to mass-produce these devices, employing the production facilities of firms making illuminated signs, musical instruments, and small mechanisms. Macrae obtained the funds to establish an engineering workshop at the Firs and brought in two "design officers" who would produce proper drawings of the rough sketches given to them and then make up the devices in the workshop. These "two characters" (Smith and Wilson—he never gives their full names), who "had no engineering qualifications whatsoever, were born designers of mechanisms." Smith and Wilson rustled up some machine tools and were soon making working mechanisms, testing them, producing detailed drawings (for production), and handing them over fully realized for field testing. This practice was widely adopted in the secret workshops. At the Frythe "there was a very well-equipped workshop there with about half a dozen hand-picked instrument makers and precision mechanics. I found that I could make freehand sketches and they would have the parts made up so quickly that very often . . . the finished article would be on my bench by the time I had come back from lunch, beautifully made.[8] This was the same process employed in Edison's famous invention factories, where skilled craftsmen like John Kruesi transformed Edison's rough sketches into working models.

Macrae realized that MD1 did not have much chance of surviving in a very crowded field as a "pirate design section," but he saw salvation in expanding into production. So MD1 built its own factory and recruited hands from nearby villages and soon began to supply mines to the cloak-and-dagger boys at Aston House at the rate of 500 at a time. By the end of 1941, MD1 was a well-equipped research and manufacturing operation with more than 250 employees and could claim to be "by royal appointment sole suppliers of nasty booby traps to the British Army."[9]

Blowing Things Up

SOE workshops produced a wide range of booby traps, explosive charges, and incendiary devices, ranging from the pocket hand grenade to the incendiary arrow. MD1 did a lot of research on shaped explosive charges for its sticky bomb—a handheld antitank weapon—and also searched for an explosive that would destroy papers in a briefcase if the agent was arrested—a difficult task because compressed paper does not burn easily. The idea of an explosive briefcase began with this in mind, and much later, the CIA carried out experiments to find the best way to burn paper quickly, but Fleming reinvented this piece of equipment as a weapon for Bond. There were several chemical laboratories experimenting with explosives: the experts at the Royal Arsenal (known as "The Shop"), the research-and-development facilities of companies like ICI, and the armament workshops of SIS, SOE, and Military Intelligence. Together they developed the practical uses of one specific type of explosive that represented a major innovation in the art of sabotage. Originated by the Shop just before the war, plastic explosive was perfect for sabotage. It came as a malleable dough that could be shaped or cut as needed. It was stable and safe—it could not be detonated when hit by a bullet—and was more powerful than all available explosives. This new explosive was to have a long and very destructive life.

Saboteurs needed a delayed-action detonator to enable them to make their getaway before the big bang. There were plenty of rope and paper fuses lit in World War II, but clandestine operations required more advanced timers that could be concealed and whose delay could be set precisely. The most widely used timing device of the war was developed by Station IX along the lines begun by the Germans in the Great War and improved by the Poles in the interwar years. A time-operated detonator in the form of a pencil-thin stick about six inches long was embedded in a primer, which in turn set off the explosive charge. This "time pencil" had a ridge on it that had to be pressed to release acid or some other corrosive agent, which ate through a wire that held back a spring-loaded detonator. Time pencils were issued in different time delays, from minutes to hours, which were indicated by a color code on the pencil. Time pencils were used in many different explosive devices from the Clam to the basic incendiary, and more than 12 million of them were produced during the war. They were also used unsuccessfully in two attempts to kill Adolf Hitler. In 1943 several high-ranking German officers planned to assassinate the Führer. Lt. Col. Rudolf von Gersdorff "decided on English plastic explosive, an English magnetic mine (a limpet), and an English chemical fuse. . . . None of the available German devices were suitable, being either too large or too conspicuous," which suggests that SOE was far ahead of its competitors in this type of equipment. Yet the time pencil was far from perfect, and Macrae of MD1 pointed out the defects of this "dodgy device": "It was so sensitive to temperature that its delay was 'almost anybody's guess,'" and in cold weather it might not go off at all. It was this sensitivity to temperature that ruined von Gersdorff's plan. The room where he intended to place the bomb was quite cold, which meant that the 10-minute time pencil he carried would take much longer to activate: "I was therefore forced to calculate on the basis on a fuse time of 15–20 minutes," but Hitler was only scheduled to be in the room for fewer than 10 minutes, so the assassination attempt was called off.[10] SOE's laboratory staff put a great deal of effort into trying to regularize the delay, including research into the chemistry of the corrosive agent and metal wire, but with little success. The Firs produced its own delay (called the L delay), which exploited the stretching properties of lead, and as it worked on mechanical rather than chemical principles, it was not affected by temperature as much as the chemical action used in time pencils. Macrae could find no factory interested in making the L delay, which required extreme precision and quality control, so he set up manufacture on-site. MD1 produced more than five million lead-delay timers—"one of the most successful war devices ever made."[11]

Communications

The need for secure communication brought the first technological innovations to espionage: codes and ciphers, the invisible inks beloved of spy fiction and Boy Scouts, and the concealment devices that intrigued Fleming. Wireless transformed secret communications, and one of the most ubiquitous pieces of spyware in the Second World War was the shortwave radio transmitter. The Soviet spy ring that operated out of Switzerland was called Die Rote Kapelle (the Red Orchestra) because of the stream of Morse code messages tapped out to Moscow every night. A fitting testament to the British lack of preparedness for the war was the outdated Mark XV wireless set, which weighed 45 pounds and came in a large plywood case, making it completely unsuited for clandestine operations. When German transmitters were captured as part of the "Double Cross" system of turning Nazi agents in England, they were examined by the experts at MI5 and the General Post Office (GPO, which had taken over some of the technical work of wireless transmission), who were impressed by their engineering and their size, a mere 12 × 6 × 4 inches. This two-valve set with its frequency adjuster was far better than anything the British had constructed.[12] These captured Abwehr sets helped Station IX to design smaller and more powerful transmitters. The type A Mk II suitcase set was 11 × 4 × 3 inches, weighed only 20 pounds, and became the most popular set in the field. The availability of compact American Loctel valves enabled even smaller sets to be produced by the end of the war.

It should come as no surprise that Fleming was much more interested in exotic weapons than in communications equipment, and James Bond rarely uses a wireless transmitter in any of the books, nor is he ever shown eavesdropping on anyone's wireless conversations. We never see him composing secret reports, struggling with transposing the letters of a code, and the only forms he mentions are his expense accounts. Yet Fleming knew all the ins and outs of communication in the spy business and describes some of them in detail in *Doctor No.* Every day at the same time, Strangways, the SIS resident in Jamaica, makes "his duty radio contact with the powerful transmitter on the roof of the building in Regent's Park that is the headquarters of the Secret Service" (9). Fleming the technological enthusiast describes the various security protocols, the call signs, frequencies, and Morse code procedures. He assures us that SIS Radio Security had the equipment and skills to detect anomalies: "The minute peculiarities in an operator's 'fist' would at once detect it wasn't Strangways at the key. . . . It was the basic protection against a Secret Service transmitter falling into enemy hands" (*DN* 13). This was certainly true in theory, but during the war Abwehr operators were able to fool the British listeners with ease and SOE operations in Holland were a complete disaster because of poor radio security. Bond certainly knows the form, too. He recognizes the sounds of the wireless transmitters as he creeps past Mr. Big's communication center in *Live and Let Die.*

In all the James Bond books and films 007 rarely sends a radio message—he uses the telephone, connecting long distance to "Universal Export" in London (the cover name for the secret service, which often hid behind commercial firms) or the CIA in Washington. When he needs some background on a suspect, he just phones up the MI6 Registry, gives his ID number and asks his question—simple enough for a superspy but firmly in the realm of fantasy. Although Bond carries a codebook, probably a one-time pad, which Le Chiffre's agents find hidden in a toilet's ball cock in *Casino Royale,* we are told little about how Bond encodes and decodes messages. He transposes his report into five-letter groups and sends them in a telegram to Universal Export in London, and in *Thunderball* he has his own cipher and cipher machine, "with the triple X setting," which he uses to transpose the text of his message before taking it off to Felix Leiter of the CIA to transmit. An Enigma-like cipher machine takes pride of place in *From Russia with Love,* and Bond is impressed with "the immensity of the prize. The Spektor! . . . Bond didn't know much about cryptography . . . but at least he knew that, in the Russian secret service, loss of the Spektor would be counted a major disaster" (83). Fleming didn't know much about cryptography either, and it hardly figures in his books, where messages are always in clear and the opposition never bothers to listen in.

Underwater

Secret operations and commando raids both faced the problem of getting their men into occupied Europe while avoiding the eyes of the enemy. Books and films about the war might lead one to believe that most were parachuted into enemy territory, but airplanes were scarce, and the RAF was desperate to hold on to them. (There were also some qualms about using their equipment to insert individuals they called "assassins" into enemy territory.) This pushed SOE into acquiring its own fleet of small boats, ranging from ancient fishing smacks to navy gunboats and modern American sub chasers. The navy's submarines and fast boats could get close to the enemy shore, but agents needed something much smaller and less conspicuous to land. All the special forces and espionage agencies used the military-issue canvas canoe and the wooden kayak for the final part of the insertion into enemy territory. This was considered such a basic task that getting in and out of dinghies and canoes in the dark was one of the subjects taught at SOE training schools. Folding canvas canoes (called folboats or cockles) were used by commandos to attack enemy vessels with limpet mines. Rubber dinghies became the preferred method of landing agents because they were lightweight and easier to stow on board ship than canoes or kayaks.

All these vessels had the disadvantage of being visible from the shore. Submersible vessels and personal equipment, such as the Davis Submerged Escape (rebreathing) Apparatus, were therefore investigated by the gadget factories. A motorized submersible canoe (the "Sleeping Beauty") was built for the Special Boat Service in 1943 to lay limpet mines. This unstable craft had some submersible properties but required its operator to wear a rebreathing apparatus, so attention turned to midget submarines. The Frythe workshop developed its own small underwater craft. The Welman one-man submersible was fewer than 20 feet long and powered by a tiny (2.5 hp) electric motor. It could move at a little more than 2 knots and dive down to 50 fathoms for short periods. Attached to the fore was an explosive charge, which was detached and placed on the hull of the enemy ship. The boat builders at the Frythe made several small submersibles, such as the four-man Welfreighter, and tested them in their water tank. This work of "enthusiastic amateurs" was done in the usual "do-it-yourself" manner without the input of the navy, which did have some experience in these matters. The Welman was not equipped with a periscope, which made it "totally useless because . . . the slightest wave made it impossible to see where you were going,"[13] and it was unstable enough to nearly drown Admiral Mountbatten on a test run in a reservoir. It was used unsuccessfully on a raid on docks at Bergen; nevertheless, it impressed the Germans enough to copy it for their midget Biber submarine.[14]

Of all the miniature submersibles constructed by SOE and the navy, the one-man human torpedo, or "Chariot," interested Fleming the most. Although not as advanced or as comfortable as the midget submarine, the Chariot demanded bravery and made warfare a personal business—one man against a great ship—and this appealed to Fleming, who always leaned toward mano a mano combat in his fiction. The Italian navy led the way in subaqua technology of miniature submarines, human torpedoes, rebreathing apparatuses, and rubber suits. The daring attack by its 10th Light Flotilla Assault Unit on HMS *Queen Elizabeth* and HMS *Valiant* in Alexandria harbor in December 1941 proved that a torpedo charge delivered by a frogman could be as decisive as one delivered by a much more expensive submarine or airplane. This attack came as a great shock to the Royal Navy, which immediately began to develop its own human torpedoes. These were employed in an SOE mission in 1942 in a failed attack on the German battleship *Tirpitz* in Trondheim Fjord. The 10th Flotilla waged a successful campaign against Allied ships in Gibraltar Harbor for three years, sinking or damaging fourteen vessels. Their campaign was mounted from a derelict tanker, called the *Olterra,* at Algeciras on the Spanish coast, just across the water from

Gibraltar. The Italians assembled the two-man human torpedoes from crates delivered to the *Olterra* and launched them from a hole cut below the waterline in the *Olterra*'s hull, returning undetected after completing their mission. This episode was a humiliation for the men of NID and not easily forgotten. Bond calls it "one of the blackest marks against intelligence during the whole war" (*TB* 143). The *Olterra* operation was copied wholescale by Fleming in *Thunderball,* where Largo's men launch their submersibles from a hatch cut in the bottom of his yacht. Fleming also borrowed some of the adventures of Commander Lionel "Buster" Crabb and his team of divers, who fought an underwater war with Italian frogmen around Gibraltar, locating and removing the Italians' limpet mines from ships in the harbor. Crabb's nickname came from the actor who played Flash Gordon, and some of his daredevil attitude might have been incorporated into the character of James Bond. In 1956 Crabb dove in a secret MI6 mission to explore and photograph the underside of the Russian cruiser *Ordzonikidze* while it was anchored at Portsmouth. This bears some resemblance to Bond's mission in *Thunderball* and probably ended in the kind of underwater fight that Fleming described, but in this case the intrepid Crabb never returned.

In 1943 some members of Fleming's own commando unit 30 AU engaged in underwater training off Algiers. They took a diving course from the submarine depot ship HMS *Maidstone,* in which they learned how to operate the Davis apparatus—a rebreather bag with an oxygen canister and a CO_2 scrubber. Italian frogmen had been entering the harbor at Algiers on two-man human torpedoes and attaching limpet mines to ships. Some were captured and their kits examined. They had tight-fitting rubber suits made by Pirelli, flippers, and breathing apparatuses. They carried limpet mines, clamps, and money belts stuffed with gear. When Italy surrendered in September of 1943, many of the 10th Flotilla went over to the Allied side. Fleming found underwater equipment fascinating and liaised with the navy's underwater experts from HMS *Vernon,* taking special interest in their one-man submersible, the X-Craft.

In 1953 Fleming joined Jacques Cousteau in diving to a Greco-Roman galley sunk in the harbor at Marseilles, and the underwater swim he took during this expedition was immediately incorporated into the manuscript of *Live and Let Die.* Cousteau had helped develop Aqua-Lung technology during the war. Instead of a rebreathing apparatus that scrubbed air clean, scuba divers used tanks of compressed air controlled by regulators. These were simpler to use and easier for an amateur to master. Cousteau's "treasure hunts and archaeological discoveries" made him into the kind of scientific adventurer who appealed to Fleming, who called him "one of my heroes." Cousteau was at the cutting edge of the science of oceanography and the inventor of several new types of underwater craft, such as a "jet propelled underwater flying saucer for submarine exploration," which greatly impressed Fleming.[15]

Much of the underwater technology developed during the war reappeared in Bond's exploits: frogman suits with a breathing apparatus of compressed-air bottles, underwater torches, small underwater craft with electric motors, motorized attachments for swimmers, and underwater explosive devices. In *Thunderball* Largo has his two-man chariots "bought, with improvements, from Ansaldo, the [Italian] firm that had originally invented the one-man submarine" (103), and supplied them to the Italian and German navies. Fleming knew the reputations of the leading manufacturers like Pirelli, used their equipment, and promoted their brand. At the end of the war he was not only thinking about writing fiction but also changing his lifestyle: "When we have won this blasted war, I am going to live in Jamaica. Just live in Jamaica and lap it up, and swim in the sea and write books."[16]

It was rather like Robin Hood and his band of outlaws . . . brave and dashing, very much like outlaws but on the side of the right.

Ed Lawson, SOE

008 Irregular Warriors

Fleming defines Bond as a secret agent. In "Risico" his duties are "espionage, and when necessary sabotage and subversion" (269). In *Thunderball* Bond tells Domino that he is "a kind of policeman" (203), and in *Doctor No* he tells Honeychile that "they send me out from London when there's something odd going on somewhere in the world that isn't anybody else's business" (92). The work that Bond carries out cannot be defined as spying, for he rarely gathers any useful intelligence that he then passes on to his control. Although some narrative elements in the Bond novels came from spy fiction, such as the chase scene and the final confrontation between hero and villain, mostly they reflect Fleming's wartime experience of irregular forces that went well beyond intelligence gathering to embrace sabotage, assassination, and other forms of justifiable violence. What Ian Fleming did in his James Bond novels was to bring the two strands of his wartime experiences together into one narrative and one hero—Bond might work for the Secret Intelligence Service, but he acts more like a commando in doing his job. So although fictional characters like Bulldog Drummond were the foundation for the Bond character, the direct inspiration came from the fighting men Fleming worked with during World War II.

Secret armies like SIS, SOE, MI9, Special Air Service Regiment, Long Range Desert Group, Robert Laycock's Layforce, and the Special Boat Section were all recruiting men determined to do "anything to get away from regimental soldiering." The military establishment considered these groups to be unruly and ill-disciplined private armies, which in many cases they were. Churchill viewed the first commando raids on Norway as the work of "a few cutthroats." Although Fleming thought these troops were "fairly piratical," they were the ideal men to fight an irregular war: independent of the bureaucracy and conservatism of the regulars, bound together with personal bonds, and depending on their own initiative and daring to accomplish things "at the sharp end." The spirit of Richard Hannay lived on in the secret intelligence services, the consequence of the popularity of Buchan's books and his part in recruiting personnel for them between the two world wars. Hannay was an inspirational figure for those in special operations who admired the daredevil, plucky, and ingenious amateur. One officer told Philip Knightley that practically everyone he met in SOE "imagined himself to be Richard Hannay." For intelligence professionals like Kim Philby, who recognized that special operations are inherently insecure ("the authors of big bangs are liable to detection"), daring but ill-considered raids were the drawbacks of the Hannay cult. Confronted with another harebrained scheme, he reflected, "Richard Hannay was with us again."[1]

Irregular warfare tended to attract a certain kind of personality: individualistic, independent (to the point of insubordination), imaginative, hyperactive, and self-reliant—all character traits prized by Fleming and exhibited by Bond. In writing about the agents of SOE, Michael Foot made the point that they liked to think of themselves as doing things differently, outside the normal codes of behavior, and "they were no longer tied to the rules of conduct in their private life," which referred to the numerous, casual sexual liaisons that were commonplace within the secret army.[2] There were plenty of opportunities to bed "the Wrens and Wracs and ATS who manned the communications and secretariat" of the secret services, as Major Smythe remembered fondly in *Octopussy* (6). These military organizations for women provided much of the operational support for SOE from office work to wireless operators, as well as some of their most famous agents. Many came from the First Aid Nursing Yeomanry (FANY, founded in 1907), including one lady "extremely glamorous in a very smart FANY uniform" who drove Lt. Tony Brooks away from the Baker Street headquarters of SOE to the airfield from which he was to fly away and parachute into France: "When we passed Hatfield she took her cap off and her long blond hair rolled down right on her shoulders. It cheered me up a bit and I thought it would be a good thing to remember when I was facing the firing squad."[3]

As a member of the SIS old-boy network, Ian Fleming met many larger-than-life characters, such as Tony Brooks, who served as models for the composite character of Bond. There was Commander Wilfred "Biffy" Dunderdale, who ran the SIS station in Paris. Biffy drove a Rolls Royce, wore the best clothes, and was popular with the ladies. He liaised with Polish émigré groups in France and maintained good relations with his French counterpart, Gustave Bertrand, who ran the French cryptanalysis unit. When Bertrand arranged for a Polish copy of the German Enigma cipher machine to be sent to Paris, it was Biffy who carried this vital cargo to London. When his train arrived at Victoria Station, Stewart Menzies (the C of MI6) was waiting for him in full evening dress, in a scene that might have been imagined by Fleming. Winterbotham saw the prize the next day in SIS headquarters.[4] SOE was full of agents who would later be called "real-life James Bonds," and several members of Fleming's own commando unit, 30 AU, have been identified as models for Bond, including Patrick Dalzel-Job: an expert marksman, skier, diver, parachutist, and pilot of midget submarines.[5] Fleming secured interesting jobs for his friends in irregular warfare, which eventually gave him raw material for the "fiction mixture" of Bond. He arranged a consulate post in Romania for his friend Merlin Minshall and the use of some naval ratings and an RAF MTB in a scheme to block the Danube. It came to nothing, and Minshall had to flee at high speed from the Germans, but his sailing skills proved useful later in France when he sailed up the Gironde estuary to see how many U-boats were entering and leaving the river mouth. A daredevil sailor, racing driver, and all-round adventurer, Minshall was described by Michael Foot as "a buccaneering type left over from the great Elizabeth's reign." During a spying mission up the Danube, a glamorous woman working for German intelligence joined him on his yacht, and the story goes that she tried to poison him, but he eluded her. Foot concluded that "certainly he shared Bond's susceptibility to blondes."[6]

SOE agents, commandos, and parachutists thought of themselves as elites and proudly carried the unique equipment that helped define their status as individuals working outside regular military organizations, which they derided as "the spit and polish boys." Parachutists were thought to be the most modern soldiers on the battlefield. They were called "airborne," "sky troopers," "shock troops," and "spearhead troopers." At the beginning of the war parachuting was considered an exotic new technology, and parachute training was reserved for elite troops; there was some consternation at the training unit in Trafford, Manchester, when some female agents of SOE turned up for the course. Parachutists stood out because of their special uniforms. The American jump jacket (or suit) and the British smock were designed to get them out of the airplane without getting snagged, and their trousers had several large cargo pockets to carry supplies into action—both still feature in fashionable wear for young men. In the British army, commandos and parachutists were the first to be issued camouflage uniforms, and Bond is dressed in "parachutists' camouflage—green, brown and black" in his assassination mission in *From a View to a Kill* (203). Jumping from great heights puts a lot of strain on the ankles; thus, American parachutists were issued calf-high, lace-up boots that were the envy of other infantrymen.

In *Octopussy* Fleming has Major Smythe put on "a pair of the excellent rubber-soled boots issued to American parachutists" (30). American parachutists were permitted to tuck, or "blouse," their trousers into their boots, and this not only made it easier to get out of the hatch of the airplane but also made them look pretty tough. When Stephen Ambrose interviewed Walter Gordon of the 506 PIR (Parachute Infantry Regiment), he was told: "We were all ready to trade our lives in order to wear these accoutrements of the Airborne."[7] So much effort went into creating special uniforms that some questioned whether it might be better to spend more time training parachutists to find their weapons on the drop zone "and rather less upon the special clothing developed for him to wear while he did the stumbling." In Evelyn Waugh's *Put Out More Flags,* one character decides to join the commandos after being seduced by their glamorous equipment: "They have special knives and Tommy-guns and knuckle dusters; they wear rope-soled shoes. . . . They carry rope ladders around their waists and files sewn in the seams of their coats to escape with."[8]

These irregular troops stood out because of their special equipment, training, and attitudes. One airborne unit was described as a "mob of pirates," and they felt that they earned the right to act like the pirates and gangsters they admired, openly reveling in their freedom to do things that would have been considered criminal in peacetime. The men who volunteered for these units were often told by their commanding officers: "Don't touch it. They're not the sort of people you want to be mixed up with." But they were exactly the sort of people who made for colorful characters in spy novels: men chosen for their toughness, self-reliance and initiative, who did not always follow orders and who thrived outside the law. One SOE trainee summed it up: "We were to be gangsters with the knowledge of gangsters but with the behavior, if possible, of gentlemen."[9] There is no better concise definition of the James Bond character than this.

The Making of an Agent

The hasty recruitment of thousands of men and women into the intelligence services necessitated an equally large training program. SOE had the task of turning civilians with a facility for foreign languages into clandestine agents, and it had to be done with indecent haste. SOE ran one of the most exclusive and intensive education programs of the war, operating a network of more than 60 training facilities across the United Kingdom and the British Empire. Commander Fleming of NID knew of these camps, had visited them, and even claimed to have passed through the course set up in the Canadian Camp X on Lake Ontario. During this visit, as an observer, he reputedly took part in all the exercises, excelling in the long underwater swim to attach a limpet mine to a target. According to Fleming's friend William Stephenson (who was in charge of the camp) Fleming was the best of all his pupils—a claim that has been contested but still forms part of the Fleming "Man of Action" myth.[10] These camps are the source for the training schools that Fleming describes in his books and that we see in every film. After Bond meets up with Q to be equipped for his mission, the two always walk through groups of men practicing with flame-throwers, machine guns, and explosive devices. We see agents running through obstacle courses and shooting at "pop-up" targets, just like they did in SOE training schools in Scotland. The opposition carried the same sort of training in the same sort of places. The Abwehr used châteaux in France, and in *From Russia with Love* Fleming describes a SPECTRE training center in a large country house where instructors mount full-scale exercises in which their chief hitman, Red Grant, can chase down and kill the victims provided for him.

SOE training was done in stages. Candidates started with a two- or three-week course that focused on physical conditioning, map reading, and shooting. Then they moved on to three or four weeks of paramilitary training in Arisaig in the highlands of Scotland, where they learned how to fire and clean a wide range of weapons, often at the end of a demanding obstacle course. Many trainees fondly remembered weeks spent in the Scottish Highlands, living outdoors, killing rabbits, and practicing dynamiting trains and bridges; others were grateful that they survived. As many as a third of the applicants might be rejected at the training schools at Arisaig, called group A. Successful candidates moved on to schools in group B, a set of country houses in and around Beaulieu in the South of England. Here they were taught how to deal with the security forces in enemy territory, how to pass through the innumerable check points and snap inspections, and how to survive interrogation if they were detained. The director of SOE's French section, Maurice Buckmaster, recalled: "They were trained . . . in self defence, in the use of arms, in the use of cover—moving around at night and so on—they must learn what the rules were in France . . . what a ration card entitled you to. . . . We also used to test them out on how they reacted under a little bit too much to drink. . . . We would wake them up in the middle of the night with a sudden bright light and see what language they exclaimed in." One trainee agent described this nighttime exercise: "About two o'clock in the morning we were woken up by batmen [servants to the officers] and mess waiters we recognised but dressed as German troops with tin hats on and rifles with bayonets. We were thrown out of bed, told to wrap ourselves up in our blankets and marched barefoot across the parade ground into the garage where Sturmführer

Folliss [their SOE trainer] was wearing his SS uniform. We were told to stand up and were harangued in broken kraut." The "somewhat rough treatment" carried out under threats and blazing lights "had all the genuine flavor. . . . Afterwards, over a whiskey-and-soda, he [or she] was told of all the mistakes he made; and how he should behave if he ever had to face the real thing."[11]

The "finishing school" at Beaulieu taught agents the rudiments of tradecraft. They learned about dead-letter drops, code words, safety signals, and how to spot if they were being followed. Students were instructed in all the different Nazi and collaborative organizations and how to recognize different types of German troops. The syllabus for the propaganda class was written by Kim Philby, late of Section D. Students received instruction into housebreaking and the handling of carrier pigeons. One student remembered Beaulieu with affection: "We learnt all the sort of things that we'd loved to have learned at school. Secret inks, codes. We had a gamekeeper come down from the (Royal) Sandringham estate who taught us how to snare and catch rabbits and things and also what to do if you were burgling a house." After the finishing school they were given special training in whatever task they were assigned to: wireless operators went to the Thame Park wireless school; others got training in explosives and sabotage at Section D's school at Brickendonbury Hall; and some took courses in forgery and safecracking. Some of the country's most celebrated criminals, like "Gentleman Johnny" Ramensky of Glasgow, were their instructors.[12]

Much of the training concentrated on survival techniques and hand-to-hand combat. During the twentieth century there were important innovations in unarmed combat imported from the Far East. SOE agents were instructed in these techniques by William E. Fairbairn and Eric A. Sykes, former inspectors in the Shanghai Municipal Police. They taught many Eastern fighting styles, including jujitsu and karate. Their hand-to-hand fighting came from the school of hard knocks on the Shanghai waterfront and appalled the military establishment as ungentlemanly "rough-house." Fairbairn took pains to reassure his critics that his techniques did not belittle the usual exercises of boxing or rugby football ("a knowledge of these is an asset to anyone intending to study all-in fighting") but that one should not shrink from "uncivilized or un-British methods" when confronted with "an utterly ruthless enemy."[13] Quite.

Fairbairn's *All-In Fighting* (first published in 1942) gives us an idea of his technique: "Edge of the hand blows are delivered with the inner (i.e., little finger) edge of the hand, fingers straight and thumb extended—the blow is delivered from a bent arm using a chopping action from the elbow." Fleming was an early enthusiastic convert to judo, which he took up as a schoolboy at Eton, and he made sure Bond knows how to fight this way: "He bent a little, and with his right hand straight and flat as a board, whipped around and inwards. He felt it thud hard into the target. The Negro screamed shrilly, like a wounded rabbit" (*LLD* 69). He kills an assassin in *Goldfinger* with "the cutting edge of his right hand . . . the weapon that had done it" (3–4). Bond is much more than a student of these techniques, for in *Goldfinger* we are informed that he intends to become an instructor and is researching his own book on the subject, called "Stay Alive," which is a compendium of all the methods of unarmed combat. Bond hopes that after completing his book, M will add it to "the short list of Service manuals which contained the tricks and techniques of Secret Intelligence" (*GF* 38).

Some of the unarmed combat techniques taught by SOE were somewhat unsophisticated. One lesson began: "The first thing you have to learn in unarmed combat is to grab your opponent by the balls." This was definitely ungentlemanly warfare. Agents who were gripped around the waist by their opponents were advised: "If possible, bite his ear. Even although not successful, this will cause him to bend forward and into a position from which you can seize his testicles with your right hand." Sergeant Harry Court told his students: "Forget any fist fights you have seen in a gangster film. . . . Never bother with your fists; it would be a pity if you damaged your knuckles by punching," and Sykes invariably ended his lectures with "kick the fellow in the balls as hard as you can."[14] There are several stories of SOE trainees killing fellow soldiers and unlucky civilians with the techniques they learned from Fairbairn and Sykes.

Narratives

Fleming's wartime association with irregular warfare was the most important influence on his novels, not only because it provided the models for the characters but also because its operations informed the plots. At the core of many of James Bond's exploits is the raid: the secret incursion into enemy territory; a reconnaissance under the shadow of darkness; a sudden, unexpected attack and then a hurried escape under fire after some sabotage is completed or valuable device stolen. Fleming's wartime activities made him part of the planning and execution of such missions. The closest he ever got to any actual combat, however, was as an observer of the Dieppe Raid, which was a major amphibious assault in daylight, but he could learn about the secret commando raids by examining the archives of NID, SIS, and SOE, from which he claims to have gotten much of the inspiration for Bond's exploits. The Bruneval Raid was probably one of the actions he studied. The clue that began this treasure hunt was an aerial photograph of a German installation on the French coast, which R. V. Jones correctly identified as a new form of German radar. A group of commandos under the command of Major John Frost were parachuted in at night, assaulted the German garrison, carried away much of the equipment (as well as one of its operators), and then fought their way to a nearby beach, where they were picked up by the Royal Navy. It brought instant validation for the concept of irregular warfare and had all the action that Fleming would put into his Bond books: a daring assault, a race against time to dismantle vital technical equipment, and a last-minute rescue by the navy.

Commando raids are the essential plot device of the Bond books and films, which often involve entering enemy territory and getting hold of high-tech equipment. First there is the briefing, usually only verbal, but *Goldfinger* has the benefit of a detailed scale model of the target, which had been an essential part of planning special operations in World War II. One subsection of Station IX went into the basement of the Natural History Museum in South Kensington and made the models for some of the most famous commando actions of the war, including the raid on the Heavy Water plant in Norway. After the briefing comes the dangerous insertion into enemy territory and a reconnaissance of the target, often carried out at night and under the noses of the guards. In every Bond film, 007 has to make the secret approach and then silently creep around enemy facilities, unseen by the guards. Fleming and Bond both think in terms of the raid, so when the latter considers the defenses around the Moonraker site, he concludes: "six tough men and all the right gear" could use kayaks to access the beach and then climb the cliffs. A tough assignment, but "if they had covering fire from a submarine or an X-craft, a good team could still do it" (*MR* 354–55).

Ian Fleming appropriated SOE's training methods, equipment, and techniques, such as using flashlights to mark out a secret airfield in *Diamonds Are Forever.* The Bond films also appropriated some of SOE's tricks. SOE once got a man safely onto a beach on the Dutch coast next to a casino by dousing him with alcohol as he left the dinghy. He then stumbled onto the beach in full evening dress and past the amused German guards. This was incorporated into the opening sequences of *Goldfinger.*[15] Some of the tricks used by SOE agents in France, such as using a coffin in a hearse for their getaway (which the Germans rarely examined), were also duplicated in the Bond films. Fleming also mined the experiences of SOE agents, as told in their autobiographies, novels, and interviews published after the war. The escape stories they told read a little like Bond: "Christiane and I tried to get out of the chateau. . . . We got out to the back terrace and sat down and pretended to make love, because we'd seen a man come around the far corner of the chateau."[16]

The escape from captivity and the villain's lair was a standard plot device in spy literature, but the experiences of the secret war enlarged this theme and focused it on interrogation and torture. Fleming's first book included an extended description of Bond's torture at the hands of Le Chiffre, which reflects the wartime experiences of SOE agents and resistance leaders in the hands of the Gestapo. Torture scenes appear so regularly in the Bond books that some critics have commented on Fleming's sadomasochistic streak, yet torture was part of an agent's experiences, and each was trained in methods to resist interrogation. Captured agents were at the mercy of a brutal enemy. Their experiences were of great interest to Fleming and Bond: "A friend of his had survived the Gestapo. He had described to Bond how he had tried to commit suicide by holding his breath. . . . He had been told by colleagues who had survived torture by the Japanese and the Germans that towards the end there came a wonderful period of warmth and languor leading to a sort of sexual twilight where pain turned to pleasure" (*GF* 141–42). Most of the outcry against the sex and violence of the Bond books was directed toward this sexualization of pain. Fleming was especially interested in inflicting pain on his heroines, who were stripped naked, spread eagled, tied down, and forced to submit to the threats and molestation of the villain and his leering accomplices. They were tortured with lighted cigarettes and left for the crabs to eat. Eleanor and Dennis Pelrine point out the frequency of bondage in the books: "Bond and friends are trussed tied and staked out for an assortment of painful pastimes."[17]

Fleming's sadomasochistic leanings might have their basis in a strain of Puritanism in his character, no doubt a relic of his stern Scottish forebears, or the brutal schooling of his childhood, in which discomfort was part of his education. The tension between his admitted life of self-indulgence and the self-mortification that reflected his "cult of personal toughness" emerged in the books in the symmetry of Bond enjoying the good life and then undergoing painful trials to redeem himself. Although committed to luxury and comfort, Bond always gives them up to bear discomfort and pain stoically. In addition to the interrogation and torture scenes, Fleming often has Bond undertake some sort of obstacle course in his escape, such as the demonic one set up by Dr. No or the underwater booby traps he encounters in *Thunderball*. Fleming's friend Selby Armitage made this perceptive evaluation: "Ian always had a frantic love of luxury. He longed for comfort, women, rich food, and expensive cars. At the same time the more he enjoyed these things, the more he seemed to need to punish himself for doing so."[18]

"Now, treasure is ticklish work; I don't like treasure voyages on any account; and I don't like them, above all, when they are secret."

Captain Smollett, *Treasure Island*

009 The Treasure Hunt

Of all Fleming's wartime experiences, his work with the intelligence-gathering "Advance Intelligence Units" (later designated Intelligence Assault Units, or AU) had the most influence on the Bond books. The idea of using irregular soldiers to penetrate enemy lines to gather secrets and new technology was originated by German Military Intelligence with the formation of "intelligence commandos." During the German parachute invasion of Crete in 1941 advance units of these commandos headed for Allied headquarters to capture intelligence personnel, documents, and equipment. In 1942 Commander Fleming sent a memo to his boss, Admiral Godfrey, that described this as "one of the most outstanding innovations in German Intelligence" and suggested that NID have its own advance intelligence unit.[1] The Admiralty thought this was a good idea, but the control of such units was the issue, and NID argued with Combined Operations and the Field Security Police until Fleming prevailed and became the point man for the organization. Every department in the Admiralty sent him lists of weapons in which they were interested, which gave him an overview of the technologies that were considered critical to the war effort. The hunt for a valuable military technology is the basis for the books *Live and Let Die, From Russia with Love,* and *Thunderball.* In the films *From Russia with Love, Thunderball, The Spy Who Loved Me, Moonraker, For Your Eyes Only, Octopussy,* and *A View to a Kill* the treasure is a cipher machine, a bomber with nuclear weapons, a nuclear submarine, a space shuttle, the "ATAC" antisubmarine device, a rogue atomic bomb, and a microchip respectively.

Nearly every English boy in the twentieth century came under the spell of Robert Louis Stevenson's *Treasure Island,* this author among them. Fleming was another whose imagination soared while reading about pirates on the Spanish Main, exotic tropical islands, secret maps, and that "Treasure Island atmosphere of excitement and conspiracy."[2] This timeless story was built on ancient accounts of the quest but stressed the centrality of information (the treasure map) and established a narrative: a description of the treasure, forming the group to find it, the journey to the place where treasure is hidden, action (betrayal, capture, escape, fighting), unlocking the clue and finding the treasure, and finally returning home richer and wiser. Bond's adventures start with a briefing from M; then there is a journey to a foreign locale and meeting up with his helpers, such as Leiter and Quarrel; then Bond mounts a raid on the villain's HQ, gets captured, is tortured, and escapes; and finally there is a grand fight scene and confrontation with the villain. Although Bond relies on air travel to start the treasure hunt, by the end of the story he is normally in the sea.

Fleming's fascination with pirates and treasure began when he was a small boy enjoying holidays on the Cornish coast. His childhood dream of finding Captain Henry Morgan's lost treasure was brought closer when he first visited Jamaica during the war and built a winter residence there afterward. "Goldeneye" was situated at Oracabessa on the north shore of the island, just a few miles down the coast from where the great Morgan installed a lookout to survey the Spanish Main. There Fleming wrote all of the Bond books. Caribbean islands figure in the plots of *Live and Let Die* and *Doctor No,* and they provided the inspiration for the name for his hero: James Bond was the author of *Birds of the West Indies,* a copy of which stood above Fleming's desk. Fleming the technological enthusiast claimed to be the first to use a metal detector to hunt for treasure, and he can list the pirates who left treasure troves and tell us what sort of coins are in these hoards (Bond recognizes them). He described some of his most sympathetic fictional characters as pirates: Darko Kerim, the head of the Istanbul station, is an "exuberant, shrewd pirate" (*FRL* 101), and the Cayman islander Quarrel, who accompanies Bond to the Isle of Surprise and Crab Key, is described as having "the blood of Cromwellian soldiers and buccaneers in him" (*DN* 32). Goldfinger's plan to seize all the gold in Fort Knox is described as "modern piracy with all the old-time trimmings. Goldfinger was sacking Fort Knox as Bloody Morgan had sacked Panama. There was no difference except that the weapons and the techniques had been brought up to date" (*GF* 181). This is exactly how Fleming updated the treasure hunt story.

Fleming certainly worked with some piratical figures during the war. He had a special interest in the naval section of SOE, which operated two private fleets for raiding along the enemy's coast from Norway to France. There he found men of Bond's ilk. The commander of the Helford Flotilla was Gerry Holdsworth, "a buccaneer, a strong character. . . . Gerry gathered round him a number of what you would call quite rough characters. . . . One of the larger-than-life characters . . . was a chap . . . who had been, I think, a sort of amateur pirate before the war."[3] The Small Scale Raiding Force was led by Gus March-Phillips, "a really rather extraordinary man. Before the war he'd had expensive tastes and slender means. He loved fox hunting and he liked driving fast cars. . . . Apart from his usual armoury of pistols and machine gun, commando knife etc., [he] always carried what appeared to be a very long cook's knife, a big carving knife. . . . He carried it down his trouser leg."[4]

Commander Fleming was a deskman who yearned to be in the field, and he planned on "pinching" Enigma machines and codebooks as soon as he joined Naval Intelligence. It appealed to his piratical instincts, and in this case the loot was worth more than gold and jewelry. His plan to crash a German bomber (several were in British hands) into the channel in September of 1940 and then ambush the rescue boat for its codebooks came to nothing. When British forces raided Dieppe in August 1942, a party of Royal Marine Commandos under Fleming's command was charged with capturing an Enigma machine (but as an intelligence insider, Fleming was not allowed to go ashore). He used this mission as the foundation for *From Russia with Love:* "The brand new Spektor machine. The thing we'd give our eyes to have. . . . To have that . . . would be a priceless victory" (83). The Allied invasion of North Africa gave Fleming's commandos another opportunity to steal an Enigma. They embarked for Algiers as a "Special Engineering Unit" and were told to look out for unusual typewriters when they were not avoiding the pro-German, French Vichy forces and units of US Army intelligence who were also after the same thing. While ransacking the offices of the German Armistice Commission, they came across a typewriter that turned out to be an Abwehr Enigma. All in all, this commando unit sent back to Britain two tons of loot, enough to justify its existence.

After "the Algiers job," 30 AU was no longer an experiment but an established commando unit. It eventually grew to three 40-man troops and a headquarters unit. It recruited tough fighting men with a knowledge of foreign languages, especially German, French, and Dutch, and included in their training were "novels and books on Intelligence work" recommended by Fleming. They had their HQ in a derelict farm in Amersham and a storehouse full of explosives, pencil fuses, grenades, knuckle dusters, knives, and weapons and silencers of all kinds.[5] Their training encompassed bomb disposal and disarming booby traps (the Germans had a nasty habit of attaching explosives to attractive-looking equipment). They also took courses in safecracking and using explosives to blow hatches and doors, and some of them were trained in photography at the Army

Film and Photographic Unit at Pinewood Studios. They used the Zeiss Contax II camera, which had a wide-angle lens (to pick up more light in dark space) to photograph the equipment they could not cart away. After D-Day Fleming's commandos went onto the continent and spent the rest of the war searching for treasure. Fleming's novella *Octopussy* has as its main character a Major Smythe of the Royal Marines, who recalls serving in a "very special task force" with affection, "haring around the country," seeking out valuable intelligence or just looking for valuables. His unit was one of six operating at the time, and was made up of "twenty men, each with a light armored car, six jeeps, a wireless truck, and three lorries" (*OCT* 21). The defeated Germans had not only military secrets to hide but also the riches of Occupied Europe that they had stolen, including the gold bars, which Smythe took for himself.

Fleming's commandos were an unruly bunch, full of initiative but lacking in discipline. He called them the "Indecent Assault Unit," and less tolerant officers often accused them of being pirates. After all, they were there to pinch things, and the line between the acquisition of enemy equipment and general looting is a fine one. Described as "armed and expert looters," they grabbed anything they could lay their hands on, and if the stuff they took was too big for their kit bags, they loaded it into stolen vehicles. Like many other soldiers in Europe, they traded on the black market and upset the American authorities by selling GIs the enemy pistols they had captured. General George Patton is alleged to have called them "a bunch of Limey pirates."[6] They got into trouble with British authorities because of their rowdy behavior and tendency to steal everything they needed. After coming out of the Mediterranean theater in 1943, Fleming told one of his officers, Sancho Glanville, to "forget anything that happened in the Med. You can't behave like Red Indians any more. You have to be a respected and disciplined unit." Playing at cowboys and Indians may be for children, but Fleming described his novels as his "Red Indian daydreams."[7] The game of "Red Indians" was also used to describe the activities of his secret agent. When Bond is captured in *Casino Royale,* Le Chiffre tells him that "the game of Red Indians is over" (140). Describing the serious business of espionage as a child's game reflects the sangfroid of the British officer class, who referred to fighting as "sport" and to a specific engagement as a "party." It also led Fleming to describe their weapons and equipment as toys.

In the last days of the war 30 AU was only one of many units looking for German secrets: R. V. Jones's scientific intelligence team reported back to SIS, the Air Ministry sent its own search party, and the army had its teams out. The French also put intelligence commandos into the field, but it was the Americans who mounted the largest effort, "Operation Paperclip," to find new weapons, research scientists, and intelligence assets. There could have been as many as 3,000 Allied military personnel involved in such operations in 1945. In comparison, there were only about 100 intelligence officers looking for Nazi war criminals, estimated to include around 70,000 suspects.[8] The Russians also carried away as much machinery and technical personnel as they could. Some were persuaded to come with offers of money; others were kidnapped. In *From Russia with Love* a Russian General Serov is in charge of the kidnapping of German atomic scientists, and in *Goldfinger* Fleming tells us that the Russians got all the German stocks of the GB nerve gas. The German intelligence archives were the highest priority of the Allied secret services because they contained the best possible evaluation of their own success and failures. With the Cold War looming, the Allies sought out the Nazi intelligence on the Soviet Union, including Russian messages that the Abwehr code breakers had broken, Russian codebooks that had been captured, and Luftwaffe aerial photographs of Russian territory. Major Smythe's mission was to obtain this information: "they kept their files all ticketty-boo. Handed them over without a murmur" (*OCT* 19). Some of these Gestapo files contained treasure. One found at Kiel dating back to the early 1930s reported that a certain young physicist named Klaus Fuchs had strong links to the Communist Party, but this was not noticed when the Allies retrieved the documents in 1945. Fuchs had left Germany when Hitler came to power and was later recruited into Britain's atomic bomb research and joined the Manhattan Project during the War, when he passed secrets to his Russian spymasters.[9]

Wonder Weapons

With the benefit of hindsight it is easy to interpret the activities of the intelligence commandos as a sort of clearing-up exercise carried out as the fighting was almost over, but at the time it had much greater importance because there were still fears that the Nazis might have a war-winning weapon up their sleeves to turn the tables at the last minute. This technological anxiety had its origins in the First World War. Speaking about the bombing of London in 1940, Fleming's friend Noel Coward reflected: "Perhaps in my adolescence during the last war, certain impressions became embedded in my subconscious mind. . . . Now . . . I am older and have watched this danger growing year by year, until we find ourselves in the tightest corner we have been in for centuries."[10] The threat from the air was still there in 1945. Allied intelligence was aware that German scientists were working on weapons of mass destruction, such as chemical and biological warfare and, most alarming, on nuclear fission: "if the Germans got that heavy water they may be able to destroy some part of London. . . . The Germans were after an explosive a thousand times more powerful as anything yet known."[11] The possibility of war-winning weapons was also stirring the imagination of the German population, who were living a medieval existence in cellars and bombed-out buildings, leaving only to forage for food and water. German civilians were comforted by "rumours . . . about secret weapons which are about to be used and will change the whole aspect of the war in our favor . . . a bomb which will be fired from a rocket from an aircraft and contains an explosive force sufficient enough to wreck a whole city."[12]

The threat of these weapons drove the Allied intelligence commandos deep into Germany. The Allied armies were followed by 30 AU into Hamburg and Kiel, two important manufacturing and research centers. The German navy had developed intelligent acoustic torpedoes that could pursue and sink Allied escort vessels. There were also rumors of an incredibly powerful engine that could make German submarines much faster than the destroyers that hunted them and did not need supplies of oxygen, which had hitherto forced U-boats to the surface to replenish their supplies. This engine was found by 30 AU in the experimental works of Helmut Walter's research laboratory at Kiel. Walter was one of the Reich's most brilliant scientists. He had designed a revolutionary jet engine, powered with hydrogen peroxide and a catalyst like sodium or calcium, that produced an explosive degeneration of superheated steam that yielded unprecedented amounts of power. These engines not only powered the Walter submarines, which could run underwater three times faster than the Type VII U-boat (the standard German oceangoing submarine), but also a jet fighter (the Me 163), which could climb faster than any other airplane and had a top speed of 600 mph, making it the fastest manned vehicle in the sky in 1945. Walter's engines also powered the V-1 missile up its ramp and provided the turbine components for the medium range V-2. The experimental facilities at Kiel were a treasure trove. Commander Jan Aylen, senior engineering officer with 30 AU at Kiel, wrote: "The average rate of finding new weapons for the first fortnight was about two per day. [Walter engine] combustion chambers were retrieved from the bottom of bomb craters. . . . At an outstation near Boseau . . . a sinister lake where midget crews, swimmers and other marine pests were trained, Walter's latest miniature 25 knot, one man U-boat was found."[13] Fleming's commandos dug up treasure in the form of buried research data, opened up safes full of microfilm, and pulled test vehicles out of the lakes where they had been scuttled. They found an alarming 16 mm film of German secret weapons, which was intended to be shown to the Führer. German secret weapons confirmed Fleming's belief in the potential of technology and stoked his anxieties about the threat of future weapons. As he looked over the reports and examined the photographs brought back by 30 AU, he must have recognized some dramatic stories in this pile of secret information.

German Science

On the eve of D-Day Allied planners started thinking about the end of the war and what would be an appropriate reparation for all the damage the Nazis had done. The Allies held German science and technology in high esteem. US State Department officials testified: "The remarkable achievements of German science before and during the present war were largely responsible for Germany's achievement of a degree of military power entirely out of relation to her population and raw material position." One piece of particularly dastardly German technology was unfortunately described by a British scientist as "a superb piece of invention. Its craftsmanship is magnificent."[14] German scientists and engineers had taken the lead in fixed and rotary winged aircraft, submersibles, missiles, jet propulsion, remote control, and chemical weapons. Any doubts that German science and engineering were not capable of delivering a war-winning weapon were dispelled on 13 June 1944, when the first V-1 missile fell on southern England. This pioneer cruise missile could outpace all but the fastest Allied fighters (the Gloster Meteor turbojets based on the Whittle engine) and carried a warhead of 1,832 pounds of amatol. As the opening shot in a barrage of missiles, sometimes more than two hundred a day, the V-1s were intended to finish the job of the 1940 Blitz and force the British to sue for peace. This terror from the skies came as a surprise and caused panic among the public and in government circles. It ended hopes that the war would soon be over. The same day that the first V-1 fell on London, another rocket came down near Kalmar, on the Baltic coast of Sweden. It was a test rocket from the German research station at Peenemunde, which had suffered a malfunction. Two Air Intelligence officers were immediately sent to Sweden to investigate the fragments of what would be called the V-2: "the multiple radio units of a complex nature . . . an Hs 293-style radio control unit . . . gyro stabilization . . . turbine driven fuel pump." It was their estimation of the capabilities of this rocket with its four-ton warhead that began the "rocket flap of July 1944." But the big scare, what R. V. Jones called "one of the biggest scares in history," came when the first V-2 fell on London, on 8 September 1944. This first ballistic missile fell at such speed that the only sign of its arrival was the double boom when it broke the sound barrier. Jones was in his office when he heard it: "That's the first one!" he told his associate Charles Frank, triumphantly vindicated after years of warning about the threat of such a weapon, often clashing with his former mentor Professor Lindemann, who did not believe that such a rocket was possible.[15]

The V-2 ballistic missiles were the ultimate terror weapons: no warning of them, no way to defend against them, and enormous destruction done at random. The jubilant German press claimed its "effect is such as to make H. G. Wells wonder whether one of his 'things to come' hasn't prematurely come to life in Germany. . . . A period of horrible and silent death has begun for Great Britain." The V-1s and V-2s realized the fears of the 1930s and made "terror from the skies" a reality. When R. V. Jones questioned the enormous amount of material and scientific resources that went into delivering a comparably small amount of high explosive, he argued that "no weapon yet produced has a comparable romantic appeal" and concluded that "one of the greatest realizations of human power is the ability to destroy at a distance."[16] This summed up the technology of terror that was to color the postwar world and inspire the fiction of Ian Fleming.

After the rocket attacks the V-1s and V-2s became top priority for Fleming's commandos, as well as every other intelligence unit. When 30 AU found a V-1 launching site at Neuilly-la-Foret, their photographs and drawings of the site were flown back to London that night.[17] Although the V-2 missile could be launched by a mobile transporter, Hitler wanted to build permanent, fortress-like structures on the French coast to house and launch hundreds of them. Fleming visited some of these installations, and he doubtless saw one of the captured missiles. His novel *Moonraker* was directly inspired by this experience. Hugo Drax's Moonraker guided missile is described exactly as a V-2: "turbines near the tail. Driven by superheated steam, made by decomposing hydrogen peroxide" (294). A Dr. Walter is the chief engineer and Drax's "right-hand man." A Nazi scientist described as the "genius of their guided missile station at Peenemunde," Dr. Walter is stereotypically German: "He sprang to attention and gave a quick nod of the head. 'Valter,' said the thin mouth above the black imperial" (312).

The postwar diaspora of German scientists and engineers spread all over the world. France, Italy, and the United Kingdom all employed former Nazis in research into rocketry and jet engines. The VfR sent its former members across the continent and to the developing military powers of South America and the Middle East. Eygpt's recruitment of hundreds of German technicians into its rocket and jet fighter programs confirmed the continued threat of Nazi science. Gamal Nasser's plans to employ Nazi-designed missiles in the 1950s earned the Egyptian president the nickname of "the new Hitler." The death rays, poison gas, and atomic bombs allegedly developed by expatriate German scientists were described in the Israeli Parliament in Tel Aviv as the final chapter of the Holocaust. One Israeli newspaper called this threat "borrowed from the adventures of Flash Gordon."[18]

In the Shadow of the Nazis

Fleming wrote his books in the aftermath of the war, and they reflect his wartime experiences. Bond sees his work as a "hangover from the war. . . . I used to dabble in that kind of thing. . . . One still thought it was fun playing Red Indians" (*GF* 11). In Fleming's books the war is a constant point of reference. He describes a "tall grim Belsen-like chimney with the plume of innocent smoke," car exhausts sounding like "an 88" (a German artillery piece), and bad weather "worse than one of those PQ convoys" (which took supplies through Arctic waters to the Russians). Fleming even reminds us that the Koreans employed by Goldfinger were also used by the Japanese as prison guards. Bond's world is peppered with former Nazis, and although the Cold War made the Russians the bad guys, the villains that Bond fights are as likely to be German as Russian. Like many other Allied intelligence officers, Fleming assumed that the Nazis would still pose a significant threat to the postwar world, and M agrees with him: "Well, we all knew there was plenty of Nazism left" (*MR* 435). The Nazis who populate the pages of Fleming's books remain the eternal enemy: "The old Hun again, always at your feet or at your throat" (HR 305). Bond confides to one lady friend, "One's had to get to know the smell of a German, and of a Russian, in my line of work" (HR 305). The villainous organization Spectre employs several men from the SS and from the Gestapo, such as Horst Ullman in *The Spy Who Loved Me*. The pilot of the smugglers' helicopter in *Diamonds Are Forever* is a Luftwaffe pilot who "fought under [Luftwaffe ace Adolf] Galland in defence of the Reich" (450), and the Italian pilot who steals the bomber in *Thunderball* also flew for the Luftwaffe.

Fleming's war work took him close to the fighting in late 1944 and early 1945, and that is why the Battle of the Bulge is often mentioned in his fiction. The SS had set up irregular "werewolf" units, which were to stay behind after the Allied invasion, and these caught Fleming's attention. Hugo Drax of *Moonraker* is first described as a British soldier seriously injured in a German attack during the Ardennes campaign, but like Long John Silver he is not who he appears to be. He is really Graf Hugo von der Drache, a fervent Nazi who began his military career in a Panzer regiment and then joined the SS and served in Otto Skorzeny's elite commando unit during the Battle of the Bulge. He was dressed in a British army uniform when wounded, and in the aftermath he was able to pass himself off as British. Sir Hugo Drax appears to be an English entrepreneur cum gentleman, and the only clue that he is on a mission of "revenge on England for what she has done to me and to my country" is that he cheats at cards (*MR* 403-7). (So does Le Chiffre, and Goldfinger cheats at golf—sure signs that they are not gentlemen.) The threat of remaining "werewolves" and of a Nazi "redoubt" in southern Germany turned out to be ephemeral; instead, war criminals had their own diaspora, which took them to South America and the Near East and kept the threat of neo-Nazism alive. Frederick Forsyth's *The Odessa File* brought them to pubic attention, along with the well-publicized work of Nazi hunters like Simon Wiesenthal. Fleming described one Nazi fugitive in *For Your Eyes Only:* Von Hammerstein was a member of the Gestapo, who looked like "the conventional Prussian officer" (*FYE* 223) and is put in charge of Fulgencio Batista's counterintelligence service in Cuba. M tells Bond that "there are a lot of Germans well dug in in these banana republics. They're Nazis who got out of the net at the end of the war" (240). Also hiding in Havana is SS Gruppenführer Sonntag.

Fleming uses Adolf Hitler as a model for his villains. Goldfinger is of small stature, and Bond doesn't trust small men, using for examples Napoleon and Hitler. Bond's archenemy Blofeld has "that lunatic, Hitler scream" (*YLT* 209). The Hitler in the Bond books and films is the Hitler in the bunker at the very end of the war, the truest articulation of the man, as the German historian Joachim Fest has argued. Encased in a secret bunker hidden underground, ranting at his subordinates, possibly demented and planning unprecedented levels of destruction, Fleming's evil masterminds share several traits with the Führer. Fleming's villains are usually men of mixed racial background who emerged from the chaotic aftermath of the war when millions of refugees had to find their way home or look for a new one. *Casino Royale*'s Le Chiffre is one of these people. A displaced person found in the Dachau concentration camp after the war, he is "a mixture of Mediterranean with Prussian or Polish strains" (25). Bond examines Goldfinger's ugly mug and decides he is "not a Latin or anything farther south. Not a Slav. Perhaps a German—no, a Balt! [someone from the Baltic states occupied by the Russians]" (*GF* 24).

Significantly, Fleming's villains often come from a background in science. Dr. Julius No is a scientist who is half German and half Chinese, describing himself as a "technician. I suit the tool to the material" (*DN* 130). Dr. No is "interested in pain. I am also interested in finding out how much the human body can endure" (148), and he intends to torture Bond and Honeychile Rider as a scientific experiment: "the facts will be noted. . . . Your deaths will have served the purposes of science" (148). Goldfinger tells Bond that he is "something of a chemist" (*GF* 103). Ernst Blofeld, who is half Polish and half Greek, studied engineering and "radionics." His partners in crime are Professor Kotze, an East German physicist ready to sell his skills to the highest bidder, and Kandinsky—a Polish electronics expert who was once head of the Philips radio research department. Hugo Drax's rocket team is "50 robotlike Germans," who display all the characteristics we expect of them: efficiency, obedience, orderliness, and devotion to their leader. Fleming writes, "Scratch a German and you find precision" (*MR* 333), but in his books scratch a villain and you invariably find a German.

Fleming clearly has little sympathy for these scientists. They are the villains with "other-wordly eyes" and little thought about the consequences of their actions. In *Thunderball* he calls them "queer fish . . . [who] saw nothing but science. Kotze couldn't visualize the risks that still had to be run. For him the turning of a few screws was the end of the job" (106). Describing Auric Goldfinger, Bond reflects, "It was the face of a thinker, perhaps a scientist, who was ruthless, sensual, stoical and tough" (23). When Fleming described the devastating bombing of Hamburg carried out by the RAF in July 1943, he noted that the city was "to the delight of the [RAF] scientists who advised on 'bombability,' terribly combustible."[19] The chief villain in *You Only Live Twice* is "a man of quite outstanding genius . . . a scientific research worker and collector probably unique in the history of the world" (83). Dr. Guntram Shatterhand travels on a Swiss passport and speaks German to his assistant, Irma Blunt. He uses his botanical knowledge to surround his castle with poisonous plants.

In Bond's world scientific research is not always altruistic. His secret agent, like his creator, houses suspicions about its dark intent: "Malignity must somewhere lie behind the benign, clinical front of this maddingly innocent research outfit!" (*OHM* 157). Bond's suspicions are proved to be correct; this laboratory was experimenting with biological warfare. Fleming's roving intelligence commandos found plenty of examples of evil science at the end of the war. In 1936 a chemist working for a subsidiary of IG Farben, the great German chemical company, tested a new organic phosphoric compound intended to be used as an insecticide. Dr. Gerhard Schrader was amazed at the potency of this compound, which was later known as Tabun (GA), and worked with his fellow chemist Otto Ambros to produce even deadlier nerve agents—Sarin (GB) in 1938 and Soman (GD) in 1944. After testing on concentration camp inmates, these poisons were put into mass production under the direction of Otto Ambros at a secret facility in Silesia. Perhaps it was only Adolf Hitler's personal aversion to poison gas that prevented the Nazis from using it after the defeat at Stalingrad. Fleming makes this devilish invention of "German chemical warfare experts" part of the criminal schemes of Goldfinger, who plans to use GB nerve gas, "perfected by the Wehrmacht in 1943" (*GF* 179), in his attack on Fort Knox.

The common stereotype of the German scientist among the Allies was that "these were brilliant men, geniuses."[20] Fleming would have agreed, but the "brilliant, neurotic Germans" he came across during the war were also capable of creating weapons of unimaginable destruction, and they seemed quite indifferent to the consequences of their actions. While Nazi chemists developed new and more toxic poisons, Nazi doctors employed human subjects for their diabolical experiments on helpless concentration camp inmates. Dr. No claims that "the German experiments on live humans during the war were a great benefit to science" (*DN* 146). The foundation of the new discipline of space medicine developed by NASA in the 1950s and 1960s came from Nazi doctors and the experimental data from the inmates they killed in concentration camps. Although many Nazi scientists argued that they were only working for the benefit of science, they had to be committed to the Nazi cause to get funding for their research, which also supported the Nazi war effort. The same Dr. Ambros involved in the development of nerve gases was also responsible for building the massive, and deadly, IG Farben factory complex at the Auschwitz concentration camp. The Allies' pressing need for scientific information usually trumped issues of morality, but there were still plenty of German scientists left to appear at the Nuremberg war crime trials. The Americans thought they had landed a prize in Walter Reppe, IG Farben's chief chemist, but the British got hold of him first, and after interrogation to obtain as much scientific information as possible, he was handed over to the war crimes court. German scientists and engineers represented only one set of technological skills developed by the Third Reich. Former members of the SS and the Gestapo were prized by developing nations for their mastery of the techniques of surveillance and torture. With assistance from the US government, war criminals like Klaus Barbie, the Gestapo officer known as "the Butcher of Lyons," were able to escape justice. Barbie, in fact, made a respectable living as an adviser to totalitarian governments in South America before he was finally extradited to France in the 1980s to stand trial.

The secret weapons that Fleming's commandos found in Germany became the basis for much of the technology of the Cold War and the anxieties that surrounded it. Fleming joined with many of his peers in the intelligence services when he described the threat of Nazi science in the Bond novels, and although Japanese and Russian scientists doubtlessly did bad things during the war, the potency of Nazi science is so great that it lives on today in the numerous films, books, and television programs about German secret weapons in the postwar world, "weapons so secretive that many remain a mystery even to this day."[21] Popular culture imagines mad Nazi scientists on the dark side of the moon, advanced Nazi airplanes that connected them with aliens and UFOs, and buried secret weapons that are still potent enough to blow up the world. There are so many feature films about Nazi zombies returning from the dead 70 years after the war that they deserve a separate subcategory of horror films.[22] Fleming's experiences of managing intelligence commandos in Nazi Germany put him in the ideal position to articulate a believable version of these fears and make the special evil of irresponsible science a central theme in the Bond novels.

"Observe, Mr. Bond, the instruments of Armageddon."

Stromberg in *The Spy Who Loved Me*

010 Nuclear Anxieties

James Bond was a child of the Cold War. Created in Fleming's postwar anxieties about the end of empire and the specter of nuclear annihilation, Secret Agent 007 had a lot more on his plate than bringing former Nazis to justice. The first Bond novel, *Casino Royale* (1953), matched him against sinister Soviet secret organizations, and the second, *Live and Let Die,* appeared in 1954, which was considered the "year of maximum danger" from a Russian missile attack, according to the US National Security Estimate NSE 68.[1] The Bond films were made at the height of the Cold War: *Dr. No* (1962) opened in the United States just after the Cuban Missile Crisis, when many in the audience were thinking about a nuclear threat emanating from an island in the Caribbean. In Berlin, American and Russian tanks were close to shooting at each other over the newly constructed Berlin Wall. *From Russia with Love* appeared right after Nikita Khrushchev was replaced by Leonid Brezhnev, who started an expansion of Russia's conventional and nuclear armament. The Bond films ran concurrently with a nuclear arms race, and a nuclear holocaust hangs over *You Only Live Twice* (1967), *The Spy Who Loved Me* (1977), *Moonraker* (1979), *Octopussy* (1983), *Tomorrow Never Dies* (1997), and *The World Is Not Enough* (1999).

The threat of atomic weapons was on Ian Fleming's mind when he first proposed a film about Bond. The story imagined the Mafia blackmailing the West for £100 million using an atomic bomb stolen from one of Britain's rocket sites. Fleming was of one mind with Winston Churchill, who spent many years worrying about the threat of nuclear weapons. Churchill was the first politician to see the strategic implications of weapons of mass destruction. In 1931 he wrote "Fifty Years Hence," about the impact of new technology from "unimaginable" methods of communication to artificial intelligence. He put special emphasis on nuclear energy: "There is no question among scientists that this gigantic source of energy exists. What is lacking is the match to set the bonfire alight."[2] A nuclear warhead atop a Nazi V-2 rocket was Fleming's first nuclear anxiety, which was aimed at London in *Moonraker.* The description of such an attack reflected his experiences of the V-weapons blitz of 1944: "The distant clang of the ambulance bells underneath a lurid black and orange sky, the stench of burning, the screams of the people still trapped in the buildings" (*MR* 436). By 1954 he considered New York to be "the fattest atomic-bomb target on the whole face of the globe" (*LLD* 7).

The historian Melvin Kranzberg spent a career drawing attention to the importance of technology in human affairs and the necessity of understanding technological change. He argued that "all history is relevant, but the history of technology is most relevant" and continually urged for a contextual approach to studying the impact of technology and its interaction with social ecology. The first and best remembered of his six "Laws" that govern the relationship of history and technology reads, "Technology is neither good nor bad. Neither is it neutral." He used examples like DDT to explain this ambiguous position, which was hailed as a wonder insecticide, a benefit to humankind, until Rachael Carson showed its devastating effect on the environment in *The Silent Spring.*[3] Kranzberg followed Lewis Mumford's argument that technology is a shaper of, and is shaped by, values and human agency. The history of atomic energy provides an example of Kranzberg's First Law as it was considered both the technological marvel of the postwar decades and, shortly afterward, the greatest ever threat to humankind. Paul Boyer has described the efforts of government and the big businesses involved in nuclear technology to persuade people that there was, in the words of a 1947 CBS documentary, "a sunny side of the atom." This campaign stretched technological enthusiasm to its limits, foreseeing a future in which nuclear power replaced all other energy sources. It was even going to power airplanes (such as the Convair X-6), ships, and a family car called the Nucleon from the Ford Motor Company. It predicted that nuclear medicine was going to cure cancer and that nuclear isotopes were going to find oil and increase agricultural productivity. With the memories of the bombing of Hiroshima receding, at least in those places not directly affected by it, the promise of atomic power was seen in terms of enriching civilization with energy too cheap to meter, boundless food resources, and a new urban civilization. Science had finally accomplished what technological enthusiasts had been predicting since the end of the nineteenth century. The output of "our great atomic research centers . . . and what it can mean for tomorrow—for peace and plenty, for health and better living everywhere." One character in *The Sunny Side of the Atom* says, "Guess folks around here won't feel so bad about that atomic bomb when they hear about this."[4]

But when the folks heard that the Russians and then the Chinese had the bomb, all optimism about a peaceful atom ended, and each of the victorious Allies threw themselves into the development of nuclear weapons, while keeping a close eye on the scientific work of the others. The military industrial complex grew rich manufacturing bombs, bombers, and rockets as postwar relations between the Allies deteriorated into a cold war founded on an expensive arms race. The policy of deterrence, in which nuclear-armed nations had enough destructive power to deter others from a first strike, maintained an unsteady peace. In 1954 US Secretary of State John Foster Dulles coined the phrase "Massive Retaliation" to deter a Soviet attack. The exponential growth of stockpiles of nuclear weapons reached the point of "Mutually Assured Destruction" (known by the appropriate acronym MAD) in the 1960s, when any nuclear exchange was guaranteed to blow up the planet. The Pentagon regularly produced war plans called Single Integrated Operational Plans (SIOP) in anticipation of a nuclear war. The SIOP-62 of 1961 designated 3,000 targets in the Soviet Union and China for nuclear attack out of the 4,000 identified in the National Strategic Target Data Base! By the time of the first Bond films, Cold War antagonists had enough explosive power to kill everyone on Earth several times over: the terror from the skies that colored science fiction and popular culture in the 1930s had been realized in a more potent form. In 1960 the editors of *Popular Mechanics* got wind of experiments being carried out on a neutron bomb, which had the advantage of killing people but leaving buildings intact. Quoting Senator Timothy Dodd of Connecticut, who called the device a "death ray," *Popular Mechanics* could claim that the "fantastic invention of science-fiction writers" had become a reality.[5]

The development of a "super" hydrogen bomb—a thermonuclear device that could yield hundreds of times the explosive power of the bombs dropped on Nagasaki and Hiroshima—increased nuclear anxieties and brought MAD much closer. Fleming and Churchill were among the millions of people appalled by the destructive power of the H-bomb, and when Churchill began his term of office as prime minister in 1951, he said that he was more worried about this threat than all the rest of his problems put together. According to Graham Farmelo this became an obsession, "a monomania," that led Churchill to hope that his final achievement would be "to lift this nuclear monster from our world."[6] Winston Churchill failed, but James Bond pulled off this feat regularly in the 1960s and 1970s.

The nuclear threat became more central in Bond's adventures as they moved from book to film. In the film version of *Doctor No* an atomic reactor powers the beams he uses to disrupt American missiles. Dr. No in the book has lost one hand to the Chinese tong he defrauded, but in the film he has lost both of them to a nuclear accident. In the film he dies as he slips into the deadly water of the reactor, but in the book he dies a much less dramatic and symbolic death, suffocating under a pile of bird dung. The atomic bomb that Goldfinger drags into the main vault in Fort Knox is the first time we see a nuclear device in the films. This is a primitive "dirty" bomb whose lingering radioactivity will destroy the value of US gold reserves in Fort Knox, but in the book it is a much smaller, "clean" atomic bomb, which is only going to blow open the door to Fort Knox's vault. (In the film Goldfinger's men use a portable laser mounted on an ambulance to accomplish this.)

Goldfinger (1964) marks the first time that Bond confronts the ultimate weapon. Here is a man who knows everything, from the best vintages of champagne to the technicalities of rocketry and radar, which he discusses quite comfortably in M's briefing in *Dr. No.* He can do the math of Goldfinger's heist in his head. He can notice things that aerial reconnaissance fails to see. He manages to master any vehicle, boat, aircraft, spaceship, and other movable equipment he encounters, from the crane in *Dr. No* to the moon buggy in *Diamonds Are Forever.* But faced with an atomic bomb in *Goldfinger,* he meets his match: he pulls and twists the cables of this ticking machine but obviously does not know how to stop the countdown clock until seven seconds (007 on the clock) before detonation, when a CIA man (formally dressed in suit, overcoat, and hat) arrives and coolly defuses it. The world has been saved from a murderous technology by a technocrat.

Fleming the technological enthusiast kept a close eye on the nuclear arms race. The bombs dropped on Hiroshima and Nagasaki were heavy and bulky and could only be dropped by four-engine bombers, which limited their employment because not every nuclear power had a four-engine strategic bomber like the delta-winged Avro Vulcan in *Thunderball.* If the bombs could be made smaller, there would be no need for a strategic bomber to carry them and thus more options for getting them to the target. A much smaller device could be the warhead on a missile, and an even smaller version might fit into an artillery projectile. In the 1959 novel the atomic warhead that Goldfinger acquires from a US Army base in Germany (for $1 million) comes from an MGM-5 Corporal tactical guided missile, a form of field artillery with a range of 75 miles first introduced by the US Army in 1955 and deployed in Germany. By the 1960s, small nuclear weapons could be fired from artillery pieces, and an immediate exchange of these weapons was expected if the Red Army attacked westward through the Fulda Gap in Germany.

The inexorable process of technological innovation in Fleming's lifetime, which took the airplane from fragile novelty to weapon of mass destruction in 50 years, increased his paranoia about the atomic threat: "Only the prototypes had been difficult . . . like machine guns or tanks. Today these were everybody's bows and arrows. Tomorrow, or the day after, the bows and arrows would be atomic bombs. . . . Soon every criminal scientist with a chemical set and some scrap iron would be doing it" (*TB* 81). In the early 1950s MI6 had gathered enough scientific intelligence to warn the prime minister that an atomic bomb could be reduced in size from the five-ton monster dropped on Nagasaki to something that could fit in a suitcase. The downsizing of the bomb made a nuclear attack carried out by a single person possible, enabling "the most deadly saboteur in the history of the world—the little man with the heavy suitcase" (*MR* 435).

Thunderball brought this threat to the fore—not in terms of a little man with a bomb but a criminal conspiracy to steal a bomb and threaten the destruction of a major American or British city if the ransom of $100 million was not paid. *Thunderball* (1965) was built around Fleming's original idea for a Bond film about a criminal organization obtaining a nuclear device, but the plot also reflects his concern about the possibility of nuclear accidents. In 1958 one of the US Air Force's newest B-47 jet bombers suffered some problems in a training flight and had to ditch its atomic bombs just off the coast of Savannah, Georgia (not North Carolina as M explains in the book). The navy mounted a large-scale effort to find the bombs, sending down frogmen and miniature submarines to search for them. This formed one of the main elements of the action of *Thunderball,* which Fleming wrote a couple of years later and a few hundred miles down the coast, in Jamaica. Captain Pedersen of the US Navy might have been speaking for Fleming when he says, "These atomic weapons are just too damned dangerous. Why any one of these little sandy cays around here could hold the whole of the United States to ransom" (*TB* 214).

Nuclear Proliferation

Taken as a whole, the Bond films provide a history of the threat of strategic bombing and an example of how the bravery and ingenuity of one man can undermine the diabolical weapons of evil organizations. This fortuitous outcome differs significantly from other films about the threat of nuclear annihilation. The "little man" with a bomb provided the plot for the British production *Seven Days to Noon* made in 1950. A deranged scientist has stolen a new, miniature bomb and threatens to explode it if nuclear weapons are not prohibited but finally sees reason at the end of the film, and his bomb is diffused. Films made in the 1960s had a much more somber tone. *On the Beach* (1960) and *Failsafe* (1964) both have tragic endings; the world is not saved by a secret agent with his gadget-laden cars and attaché cases. Stanley Kubrick's *Dr. Strangelove or: How I Learned to Stop Worrying and Love the Bomb* (1964) depicts the end of the world, made poignant by the soundtrack of Vera Lynn's "We'll Meet Again," one of the most popular songs of World War II. Kubrick's mad scientist could well have been a Bond villain. A good Nazi, faithful to the Führer, a thick German accent (copied from Professor Henry Kissinger), and a scientist who was physically and morally damaged, Dr. Strangelove is a technocrat without a conscience and should have been an extremely threatening character if Peter Sellers had not made him so comic. Yet the funniest scene in the movie, and the best line—"Mein Führer, I can walk!"—still represents technology out of control in the form of an artificial hand with a mind of its own. Kubrick had intended to make a serious film about the nuclear threat, but after he researched it, he decided that the nuclear arms race was so absurdly dangerous that only black humor could do it justice.[7]

The research that went into the Bond films diligently tracked the development of the nuclear threat without passing judgment on the new technologies of death and destruction. The films depict deadly machines from jet bombers to ICBMs with the same reverent gaze that admires Aston Martins and bottles of expensive champagne. True to the aesthetic of Ian Fleming, who saw in the V-2 "the terrible beauty of the greatest weapon on earth" (*MR* 317), the Bond films find beauty in weapons of mass destruction; the bombs, rockets, and submarines are center stage and brilliantly illuminated so that we the audience can wonder at their power and majesty.

The sleek, streamlined nuclear submarine entered popular culture in the 1950s when the USS *Nautilus* traversed the North Pole underwater in a widely publicized promotional cruise. Appropriately named after Captain Nemo's futuristic vessel in Jules Verne's *Twenty Thousand Leagues under the Sea,* the *Nautilus* brought science fact to science fiction with its astounding speed (it was far faster than other submarines) and its ability to stay underwater until its supplies of food ran out. The *Nautilus* was a direct descendant of the German XXI U-boats that had been introduced during the last months of the war and had been captured by units like 30 AU and brought to British and American naval yards. The US Navy's first postwar submarines were designed around this German technology, and the *Nautilus* class was an enlargement of the design to fit a nuclear reactor built by Westinghouse. The reduction in size of missile guidance systems and the increased yield of small warheads allowed the navy and its civilian contractors to build a missile that could be fired from such a submarine. Test firings began in the late 1950s, and the Polaris A1 was made operational in 1960. This was a solid-fuel missile launched from under the surface, which made it much less vulnerable than a bomb carried by an aircraft (which could be shot down) or land-based ICBMs, which were all targeted by Soviet missiles and could be expected to be eliminated in a first strike. The nuclear-armed submarines that prowled the depths were difficult to locate, and their complement of missiles, each of which had numerous independently targeted warheads, could eliminate all the major cities of the East Coast of the United States in just one attack. The American, British, and Soviet navies quickly built nuclear submarines, and the Royal Navy's Polaris fleet, stationed at Holy Loch in Scotland, was on the front line of NATO's defense against Soviet aggression.

Up to this point the Bond novels and films had imagined the threat coming from the skies, and Bond had overcome it by taking to the sea, which he did in *Dr. No, Thunderball,* and *The Spy Who Loved Me.* The Polaris-armed nuclear submarine moved the threat from space to sea, and this turned the focus of military research onto submarine detection, and the attention of the Bond film producers Broccoli and Saltzman to underwater theatrics. This not only brought the latest and most deadly technology to the Bond films; it also discouraged the competition from copying the underwater action because it was so expensive to film. *The Spy Who Loved Me* took Bond back underwater in a story that resembled *Thunderball,* but instead of the jet bomber armed with atomic bombs, the target of the hijackers was nuclear submarines armed with Polaris ICBMs. The film begins with some location shots of the Polaris fleet at Holy Loch. True to their claim to verisimilitude, the set designers were allowed a peek inside a nuclear submarine. Nevertheless, the interior built for the film was more than twice the size of any military submarine and far more luxurious, but as production designer Ken Adam pointed out, the Bond films had to be larger than life.

As nuclear proliferation brought many more nonaligned countries into the nuclear club and made weapons more available in the 1980s, atomic anxiety in James Bond's world continued with *Octopussy,* which has as its nuclear threat a renegade criminal organization obtaining an atomic bomb. In *Thunderball* the nuclear weapons at the center of the plot are depicted in the traditional shape of bombs, with fins and a streamlined shape, like the "Fat Man" or "Little Boy" bombs dropped on Japan. In *Goldfinger* the bomb is depicted as a piece of machinery contained in a sleek aluminum case, about the size of the deep freezer units being sold to American households in the 1960s. The special effects team who built it for the film were told only to make it look modern and active, with wheels spinning and a ticking digital clock rather than a silent, sinister device. Goldfinger's bomb was designed to look alive and working, and its clean, aluminum body gave it a look of sleek efficiency. By the time of *Octopussy* the look of the bomb has been transformed. No longer sleek and futuristic in its design, it is now functional and painted a drab khaki. The size of a beer keg, it comes in a metal ammunition case slung between two poles and can be easily carried by two men. The Bond films chart the evolution of the atomic bomb from ominous projectile to utilitarian luggage. According to the film's production designer, Peter Lamont, the bomb had to appear to be authentic.[8] It is armed by a cylindrical detonator that has two LED displays (one counting down the time to detonation), a dull red glazed panel at the top (like the sensor that picks up the signals from a remote control), and some Russian Cyrillic characters printed horizontally across the face of the device. The front of the detonator looks like a large version of the face of the Seiko watch that Bond consults as the time runs down. After insertion into the body of the bomb, the detonator is armed by a complicated series of actions—"set the time for the explosion here . . . and twist the lever a quarter turn clockwise"—but instead of the hesitation of the 1960s, the 1980s Bond knows exactly what to do and removes the detonator at the last moment.

The Bond films articulated technological anxieties in the way they depicted nuclear weapons, and over the course of the series they educated audiences in the technicalities of the nuclear threat. Viewers were made aware of the differences between weapons-grade plutonium and fuel-grade plutonium (used in reactors), knew the ominous meaning of the word *megaton,* and were accustomed to watching the transfer of radioactive materials with robotic arms. By the time of *The World Is Not Enough,* audiences had been brought up to date about the programs to reduce stockpiles of atomic bombs. Some of the action occurs at a decommissioning site in the old Soviet Union, where Bond again has to prevent the bad guys from stealing a bomb. Audiences also saw a wide variety of nuclear weapons in the Bond films. The bombs in the Russian site are all shaped like projectiles and come in shades of military khaki, but the device that Bond must defuse is a much more complicated and dramatic-looking device; shaped like a satellite, with a metallic, spherical head, and filled with a mass of wires, this silvery, futuristic device owes a lot to the aesthetic of space travel. The millennium Bond requires some expert scientific assistance, provided by the beautiful Dr. Christmas Jones, to diffuse this bomb. The climactic fight scene of the film occurs in the reactor room of a nuclear submarine, where Bond has to prevent the villain from inserting a plutonium fuel rod, which looks like a golden phallus, into the reactor and causing an explosive chain reaction in a kind of machine-driven orgasm.

By the end of the twentieth century the Cold War was over, but the nuclear threat existed in a more diverse form of criminals, madmen, and terrorists capable of building and detonating an atomic bomb. This provided more grist for Eon Productions' mill and was wickedly satirized in *Austin Powers: International Man of Mystery* (1997). After Dr. Evil's ideas for criminal enterprises have all been rejected by his associates, he sighs and says, "Okay, we'll do what we always do—highjack some nuclear weapons and hold the world hostage."

Weapons of Mass Destruction

Fleming and Broccoli and Saltzman exploited nuclear anxieties to raise the stakes of Bond's adventures, but the same pressures that made them devise more ingenious gadgets forced them to create more menacing threats to civilization. Chemical and biological weapons of mass destruction were technologies perfected in World War II, and Fleming's involvement in the hunt for German scientists in the last days of the war brought him face-to-face with a weapon that he considered as insidious and terrifying as the atomic bomb. German chemists had developed a new form of poison called nerve gases. Rather than attacking the lungs, these substances infiltrated the victim's nervous system, blocking the nerve junctions and thus preventing the brain from controlling the body. A drop of this agent could kill hundreds of people by merely coming into contact with the skin, and it was far more lethal than anything used in either of the world wars. One V-2 payload of nerve gas could have killed most of the population of London. Adolf Hitler recognized the potency of this toxin and ordered IG Farben to construct a factory to produce a thousand tons of Tabun a month. It went into production in 1942 and was captured by the Red Army in 1945.[9]

Fleming had gone beyond the threat of nerve gas by the time he wrote the Bond books. He gives a detailed explanation of chemical and biological warfare in *On Her Majesty's Secret Service* in the chapter titled "Something Called 'BW.'" Fleming spends many pages educating the reader on its chemistry and destructive potential: "We talk about the new nerve gases the Germans invented in the war. We march and counter-march about radiation and the atomic bomb," but a biological attack can devastate "thousands of square miles" (*OHM* 233). This weapon is especially menacing because it does not need expensive delivery systems like rockets. Fleming's expert from the ministry tells us: "The nature of BW agents makes them very adaptable for covert or undercover demonstrations" (233). Bond has to face World War I–era poison gas in *You Only Live Twice* and *The Spy Who Loved Me,* but the nerve gas GB that Goldfinger plans to use is described by Fleming as "a more effective instrument of destruction than the hydrogen bomb" (*GF* 179). Although *On Her Majesty's Secret Service* was intended to be a film free of gadgets, avoiding the excesses of its predecessors, the threat of perverted science was just as great. In the book the villain Blofeld plans to interfere with British agriculture, but in the film his "virus omega" is going to produce infertility in all plants and living things, including humans. The film raises the stakes from the decline of English turkeys to the survival of life on planet Earth. Blofeld plans to spread his viruses by concealing them in cosmetics carried by his brainwashed patients far and wide—a suitably frugal delivery system for a film that was low budget in relation to the other Bond films. Ten years later in *Moonraker*—a much more expensive film—the villain delivers his toxic agents from space in satellite-like pods that Bond must destroy with laser beams fired from the space shuttle—just another day at the office for him.

These weapons of mass destruction give the bad guys "the power to reshape the world" (as one of them says in *The World Is Not Enough*), but it also gives Bond the opportunity to save it. The adventure films of World War II had put the lives of hundreds and then thousands at risk, but in the Cold War the stakes were raised to hundreds of thousands and then millions of victims. What better threat could a secret agent, one man with a gun, guard us against? Bond still carries out the heroic commando raids that so impressed his creator during the war, but the disabling of weapons of mass destruction and his willingness to sacrifice his own life if necessary elevate Bond to near messianic status: "It's me or a million people" (*MR* 416).

He [Fleming] found an outlet for his passions, putting all his fantasies of supermen, pistols, women and cars onto the shoulders of Bond.

Kim Philby, *The Philby Files*

011 Gadgets

Although Ian Fleming had described Bond's spyware expertly in the books and incorporated them into the plots, the films did not initially follow this practice. *Dr. No* is a film with few gadgets, and the spyware is basic: talcum powder on the locks of his briefcase, a hair on cupboard doors, breathing tubes for underwater concealment, and cyanide poison in cigarettes. There were no modified cars in *Dr. No.* Bond drives a Sunbeam Alpine, a sporty two-seater, in both book and film. The Alpine was a production model with no modifications from Q branch, but it was still too noticeable for Bond in the book, so he has Quarrel replace it with a more sedate family car, an Austin A30, which Bond would never be seen driving in the films. Much of the equipment in *Dr. No* reflects the heritage of World War II. Dr. No's guards carry Sten guns and wear khaki army uniforms with Sam Brown belts and English army–style canvas holsters. The scene in which Mary Trueblood is killed while operating her radio transmitter was commonplace in war films—the fate of the many wireless operators of the French Resistance. MI6's wireless room reflects the look of SOE's communications centers yet with far more modern equipment, which the film's set designers borrowed from an English company equipping the control tower of New Delhi airport.

The most useful gadget in Bond's arsenal is the Geiger counter, which he uses to track down the radioactivity that leads him to Dr. No and uncovers his evil scheme. This instrument for measuring radioactivity was introduced in 1908 as the Geiger-Muller device. The first portable Geiger counters were introduced in the 1950s, and it is one of these units that we see in the film. Small enough to be portable, with clearly displayed dial (the first models only made clicking noises picked up on headphones) and large, easily managed switches, this plastic-encased instrument was easy to operate. Bond tests it by running his watch (with its radioactive luminous dial) over it. Nuclear anxieties in the United States led to a government-sponsored building program of underground shelters and the equipping of the civil defense force with thousands of Geiger counters. What had once been an advanced testing instrument was now a piece of high-tech equipment appropriate for a secret agent in the nuclear age.

Despite lacking modern gadgets, *Dr. No* reflects 1960s modernism, with its innovative titles, crisp Technicolor, and fast cutting. The director Terence Young admitted that it wasn't much more than a *Boy's Own* adventure story, but it had glamour, excitement, and Hollywood production values: "It was really the first conscious American picture made in England."[1] Ken Adam said that he set out to design a film for the electronic age, which was inspired by the angular and functional 1960s aesthetic and the look of new machines like computers.[2] In the book Dr. No's subterranean headquarters is a combination library/study with books to the ceiling and comfortable club chairs in the British gentleman's club tradition. In Adam's film set the bookshelves are replaced by rough stone walls, which reflected California-1960s modern houses, and a huge floor-to-ceiling aquarium window—a scaling up of Captain Nemo's quarters in Disney's *20,000 Leagues under the Sea.* The spaciousness of Dr. No's lair reflects 1960s interior design, but it also represents the grandiosity of the villain's ambitions. As Cristoph Lindner points out, "Large-scale international crime requires an equally large-scale base of operations," and this dictum was followed faithfully by Eon Productions. Dr. No's headquarters has the same holding cells, conference rooms, and torture chambers found in any decent Nazi establishment, but Fleming points out that it is "the most valuable technical intelligence centre in the world."[3] This combination space-launch/nuclear-reactor control center is crammed with instruments and flickering indicator boards. The mechanical arms grasping distilled evil behind glass walls and the hazmat costumes of the laboratory workers working them was probably inspired by the production staff's visit to the Harwell atomic research facility in the United Kingdom. The equipment in Dr. No's control room reflects the modernization of instrumentation; there are banks of lights and television monitors of all sizes, but the wheel that Bond turns to sabotage the reactor, with a large sign above it that reads DANGEROUS LEVEL, reminds one of *Metropolis.*

From Russia with Love is a Cold War drama in which the Russians are the bad guys, but the film remains within the technological context of World War II. The plot revolves around an Enigma-like code machine that can be carried by hand, and this accommodates a fast-paced treasure hunt (where a much more appropriate but bulky IBM punched-card computer would not). In the book

From Russia with Love Q branch provides Bond with a Swaine and Adeney leather briefcase containing ammunition for his Beretta, throwing knives, gold sovereigns, a silencer for the Beretta hidden in a tube of Palmolive shaving cream, and a "cyanide death-pill," which Bond immediately throws away. These are "the tools of his trade" and not much different from the equipment that SOE agents took with them during the war, including the standard issue "L" pill. But in the film Q has added a booby trap to the briefcase, a canister of tear gas hidden in a tin of talc. It is fixed to the inside of the case and activated if the locks are opened by moving the catches outward in the normal way. To avoid detonating the bomb inside, one has to turn the catches 90 degrees—from vertical to horizontal and then move them outward to open the lock. This is the device Bond uses to turn the tables on KGB agent Red Grant, whereas in the book his metal cigarette case stops the bullet fired at him—saving his life as it did for several lucky soldiers in two world wars. The attaché case also contains an advanced ArmaLite rifle, establishing Bond as a purveyor of the latest weaponry. The films constantly reflect advanced technology in Bond's equipment, burnishing Fleming's World War II gadgets with a 1960s aura of modernity.

When the producers canvassed the audiences of *From Russia with Love,* they found that they remembered all the gadgets and liked the character of Major Boothroyd, the quartermaster. Bond gets his equipment from Boothroyd in the book *Doctor No,* and although Fleming often refers to Q and Q branch, they are backroom boys who play no significant part in the novels. The films created the character of Q, and played by the actor Desmond Llewelyn, Q became the most popular of all the supporting roles. After *From Russia with Love* Broccoli and Saltzman hired the same actor to play Q, and they put their production designers and special effects team to work on producing more photogenic spyware for him to issue to Bond. Equipping 007 became an obligatory part of the Bond films from then onward.

The spectacular success of *Goldfinger* proved that Broccoli and Saltzman had been right to make Q and his gadgets an integral part of Bond's world, and one particular gadget captured the public's attention and became permanently associated with James Bond. The heavily modified Aston Martin DB5 appears in *Goldfinger* for only 13 minutes, but it has claims to be "the most famous car in the world" and is the subject of magazine articles, documentaries, and adoring internet sites. Two DB5s were used in the film, and the one that Connery drove was last valued at $4 million (before it was stolen), making it one of the most valuable automobiles ever built. The automobile as a handmade accessory for Edwardian adventurers was represented by Bond's racing Bentleys, but the

Aston Martin company picked up this baton after the war. Founded in 1913 by two technological enthusiasts, Lionel Martin and Robert Bamford, Aston Martin custom-built fast cars for gentlemen to race. They entered the Grand Prix circuit in 1922, and Fleming undoubtedly watched some of their cars roaring around the Brooklands track. Fleming replaced Bond's vintage Bentley with an Aston Martin DB3 in the book of *Goldfinger,* and although the DB3 was very fast, it looked functional as it was purely a racing machine. Its successor the DB4 was styled in Italy, and its graceful shape and luxurious curves moved the brand from stripped-down racing car to Grand Tourisimo. The DB5 followed suit and was acclaimed for its stylish looks when it debuted at the 1963 Earls Court Motor Show. It was also exclusive—a few more than a thousand were made. Ken Adam wanted to give Bond something more modern and fashionable than the old "green label" Bentley he drives in the books: "We decided the Aston Martin DB5, the most expensive and sexy sports car of the period, would be the right prop for Bond."[4] It was. James Bond would not only be seen in beautiful locations with beautiful women but also with beautiful objects. While Fleming concentrated on function, the attention of Eon Productions was always on form.

Comparing the Aston Martins of the films with the Bentleys of the books provides an insight into Fleming's attitudes toward technology in general and specifically toward Bond's equipment. His Bentley is all power, and the modifications—such as twin oversized exhaust pipes—are intended to make it go faster rather than make it look attractive. The Bentley's body is a simple two-seater with no added comforts, and Fleming takes the time to tell us that it is painted "in rough, not gloss, battleship grey" and that Bond has replaced the distinctive Bentley "winged B" hood ornament with a utilitarian "big octagonal silver bolt" (*TB* 72). In contrast, the DB5 is all style, and its Silver Birch paint gives it a handsome gloss to emphasize its sleek lines. Adding machine guns and an ejector seat completed the fusion of modernity and style essential to Bond's equipment. In the book *Goldfinger* the Aston Martin has switches to turn off the front and back lights independently (to deceive followers at night), reinforced steel bumpers (to ram an opponent), a radio location device called a "Homer," and a concealed long-barreled Colt 45 "in a trick compartment." In the film it came equipped with an array of cleverly concealed weapons. Most of the ideas for weaponizing the car came from the war, such as the ejector seat, which had been developed by the Luftwaffe and was now a recognizable part of the equipment of jet pilots. The oil slicks, smoke dispensers, and tire destroyers were all old ideas brought back from France by SOE agents and incorporated into the books, but they were now activated by electrical switches in the driver's console rather than mechanical levers. Bond can engage the electric motor that turns the revolving number plate from within the car. The machine guns installed in the front of the DB5 were logical modifications desired by many motorists at one time or another.

Goldfinger cost $3.5 million to make and returned $23 million, making it one of the most profitable films ever made up to that time. It earned back its production costs in its first two weeks on only 64 screens and became the highest grossing film in the United States in 1965. Its success at the box office and a worldwide marketing campaign staged by Eon Productions marked the beginning of "Bondmania," a global surge of interest and affection that rivaled the public adulation around the Beatles, who were captivating audiences around the world while cinemagoers were lining up to see Bond in *Goldfinger.* Penelope Gilliatt's review in the *Observer* concluded: "*Goldfinger* belongs absolutely to our period. So does the command of technology, the stylish brutality, the wit and the nerveless treachery." The film historian David Thomson wrote that "it was plain in the 1960s that Britain was greedy for anything modern" and that this modernism was the essential difference between the films and the "rather brutal, old-fashioned books."[5] This modernity was suggested by the Aston Martin and spyware that reflects the technology of the future. Fleming described several homing devices in the books that work with vacuum tubes and dry batteries, and Bond has to listen to the signal, watching the decibel levels and not allowing the noise to fade: "a simple form of direction finding." But in *Goldfinger* the clumsy vacuum tubes have been replaced by transistors, which reduce the size of the equipment, allowing Bond to place a signal generator into the sole of his shoe. In the film the homing device has a visual display that sits behind the cover of the car radio's speaker. It is in the form of the moving map, with the signal pinpointed over a map of the area—a make-believe gadget that preceded GPS in cars by 30 years.

The Bond films also highlight the advances in sound-recording technology, which had exploited innovations such as magnetic tape developed during the war. M and Miss Moneypenny have tape recorders on their desks in the films and 00 agents are equipped with smaller models. In *From Russia with Love* a quarter-inch tape recorder is concealed in a Rolleiflex camera, and in *Thunderball* a book encases a reel-to-reel sound-activated tape recorder. Fleming had previously taken Bond from the era of acetate discs and wire recorders to the convenience and high fidelity of magnetic tape, and the films followed this forward into the era of cassette tapes, which form the technological MacGuffin in *Diamonds Are Forever* and *A View to a Kill.* In contrast, Bond's real-life employers continued to use outdated equipment in their surveillance; some of the interrogations of the suspected spy Kim Philby in the 1950s were recorded on the old acetate discs.[6]

Broccoli and Saltzman were constantly on the lookout for visually dramatic equipment, and this eventually led them to the frontiers of science. The scene in which Goldfinger threatens Bond with a laser beam is one of the most famous in all the Bond films. This was the first time that film audiences had seen such a futuristic device (readers of the book had to be content with a circular saw). The producers had a real laser brought into the studio to see how it looked, but the beam of light it produced was too slight to be noticed in the bright lights of the studio, so the special effects people built a laser that resembled the death rays of science fiction and added the light beam in postproduction. The effect of the laser cutting through the table was achieved by using an acetylene torch under it—leading to one of the few moments in the Bond series where 007 really breaks out in a sweat.

In each successive film Broccoli and Saltzman had to contend with the same problem faced by Fleming as he started each new book: how to keep ahead of the competition by maintaining a sense of authenticity while providing thrills, spills, and even more fantastic gadgets to keep the audience's attention. The Bond films were so profitable that several Hollywood studios quickly produced their own secret-agent films. As returns from Bond films grew larger, more spy films were put into production—at least 20 were in the works in 1964, and this number rose quickly to 50. More than 60 spy-themed films appeared in 1966, from *Torn Curtain,* by Alfred Hitchcock, one of the several directors who were drawn to a more realistic Bond, to the exploitative whimsy of *Dr. Goldfoot and the Bikini Machine,* in which the mad scientist Dr. Goldfoot (played by Vincent Price) invents sexy female robots to seduce the wealthy and powerful and thus take over the world. Even the Beatles got into the spy act with their second film, *Help!*

Equipment played an important part in the spy films released during Bondmania. The action-adventure films that duplicated the heroic and oversexed secret agent often took gadgetry to the limits of credibility. Derek Flint is a superspy in every sense of the word: a master of martial arts, a pilot of his own Lear jet, a gourmet who knows his bouillabaisse, and a man who enjoys his women four at a time. He takes Bond's self-confidence and savoir faire to unprecedented and often annoying heights, and *Our Man Flint* (1966) went one better than Bond in practically every category. Flint's cigarette lighter had "82 death-dealing devices" installed in it and could light cigarettes, as well. The villains in *Our Man Flint* are a trio of white-coated scientists, Drs. Schneider, Woo, and

Krupov (representing the nationalities of the Holy Trinity of Cold War villains), whose lair is an island hideaway. They acknowledge with envy "the American fondness for marvelous toys," and Derek Flint fights them with the standard equipment of 1960s spy films: miniature radios, Geiger counters, microscopes, and multipurpose watches. Flint operates in an environment of microphone bugs, remote CCTV cameras, miniature radios, and a red telephone connecting the caller to the highest levels of the chain of command. Matt Helm in *The Silencers* (1966) has a bachelor pad that only the readers of *Playboy* could imagine, complete with circular bed that tips the hero into a large bathtub full of scented water and the obligatory blond. In this modern lifestyle everything is automated, from drying down after a bath to making coffee. *Get Smart* (1965–70) brought the absurdity of Bondian gadgets to television and played them for comic effect, fitting telephones into more than 50 props, including a clock, comb, tie, planter, headboard of a bed, cheese sandwich, lab test tube, Bunsen burner, and, most notably, in one of Maxwell Smart's shoes, which he has to take off in order to receive calls. Agent 86 and his female sidekick, Agent 99, are also equipped with a variety of weaponized telephones. Absurd gadgets were especially suited to the cartoonist and found imaginative variations in the "Spy vs. Spy" strips of *Mad* magazine.

Any spy film that hoped to compete with James Bond had to have more gadgets and bigger sets. There were, of course, the elaborate conference rooms and hideouts on remote islands or in volcanoes, but the most important were the high-tech control centers from which the villains directed their criminal activities. In the 1960s one of the symbols of high technology was the large, self-standing computer, like the IBM 360. These went into government, big businesses, and university laboratories and soon became obligatory equipment in sci-fi films, replacing the meters and chemical retorts found in the lairs of earlier mad scientists in silent films. The computers in 1960s films always come up with the answers. Derek Flint is chosen for his mission not by an American version of M but by a large card sorter that feeds on punch cards and spits out his name. Matt Helm's name pops out of an automated Rolodex connected to a television monitor in *The Silencers.* (In these early days of computers their output comes on pieces of paper.) Some films, like the Bond series, aimed at realism and modernity; others leaned toward more imaginative and colorful props in which the instruments and computers were just painted on the set—bright colors, weird dials, and blinking lights.

Broccoli and Saltzman were adamant that their spy films reflected the technology of modern espionage and stressed the effort that went into ensuring realism, but Broccoli pointed out, "With each new Bond picture we *have* to be bigger, better, more spectacular, more exciting, more surprising than the previous ones. Dreaming up new stunts, new twists, original gimmicks . . ."[7] Bond operates the most modern cars, helicopters, aircraft, hovercraft, and submarines, and they all come with the latest weapons systems. All his Aston Martins are armed to the teeth. The Lotus Esprit he drives in *The Spy Who Loved Me* has rear-firing cement and ink ejectors, and the autogiro he flies in *You Only Live Twice* is equipped with guided missiles. As his equipment becomes more deadly and more futuristic with each film, it moves from the real to the imagined. The latest equipment of the military industrial complex was fashioned into even more futuristic props built by artisans to the whims of the film's designers. This process had begun in *Goldfinger,* for if one knew anything about the weight of ejector seats and the physics of explosive propulsion, then the Aston Martin DB5 was too lightweight to be able to contain both ejector seat and its explosive charge. As firing a real ejector seat would have destroyed the chassis, the props men used a dummy ejected out of the car by compressed air.

At the same time, the Bond films did contain some real equipment that represented cutting-edge technology. The rocket belt that Bond employs in the opening credits of *Thunderball* had been anticipated in the 1930s by film serials and adventure stories when the hero is strapped into a personal jet-propulsion device. By the 1960s the Bell Aircraft Corporation had produced a jet pack for the US military, and one of Broccoli's friends in uniform told him about this secret project. It came to the film set fully operational and with its own pilot (who refused to fly it without a safety helmet, forcing the producers to film Connery in an unheroic helmet, too). It never became standard equipment in the military because it was far too unstable and its post-Bond career was in the entertainment industry. At the end of *Thunderball* an airplane lifts Bond and the girl to safety with an amazing hooklike device. The US Airforce's Sky Hook rescue system was a real-life piece of technology that grew out of a pressing need to extract valuable personnel from remote locations in a hurry. The US Postal Service had used aircraft to pick up bags of mail in the 1930s with a device invented by Lytle S. Brown, and this was developed by All American Aviation and used in 1943 to make the first human extractions, but the force of up to 17 Gs proved to be dangerous and often injured the person dangling at the end of the line. Toward the end of World War II Robert Edison Fulton was busy building a flying automobile when the CIA got him interested in an Aerial Retrieval System. Fulton realigned the pickup hooks at the front of the aircraft (to give the pilot a better chance of engaging them) and used an ingenious pulley system to pick up the slack and avoid injuring the agent as he was grabbed. The CIA sponsored the development of Fulton's idea, and it was used in Operation Cold-Feet in 1961, in which agents were parachuted into a Soviet weather station in the Arctic and sky-hooked out, along with a treasure trove of codebooks and scientific equipment.[8]

Bond and the Life Aquatic

Thunderball had several underwater scenes, and although Broccoli and Saltzman recognized that taking the Cold War underwater was going to be vastly more expensive than filming on land (the underwater action scenes cost well over $1 million), they were already committed to "bigger and better." *Thunderball*'s record-breaking revenues ensured that the frogman exploits that inspired Fleming during the war became part of the Bond formula, and the large water tank at Pinewood Studios saw years of service helping to recreate Bond's underwater missions. Bond always dives with the latest gear, but some of his equipment was made up by the props personnel. The underwater jet pack that he carries on his back like an ordinary Aqua-Lung has a jet-powered propulsion unit enabling him to race about underwater. It is fully weaponized with harpoon guns, a searchlight, and a smoke-screen generator. This fantastic device was designed by Jordan Klein, a pioneer in underwater photography who won a special effects Oscar for his work on the props of *Thunderball*. The plans of the sea sleds that transport Largo's men and the stolen bomb were sketched by Ken Adam and sent to Klein, who then built working models. Much to Adam's surprise they actually worked underwater after a few modifications. This technology had first been imagined by Jacques Cousteau in the 1950s, and today underwater propulsion vehicles are sold by the SeaDoo and Stidd companies for recreational use.

For Your Eyes Only was another underwater treasure hunt that featured advanced diving equipment. The "Jim Diving Suit" was an atmospheric diving suit named after Jim Jarratt, who used one to locate the wreck of the *Lusitania* in 1935. It was used during World War II and gradually improved so that by the 1980s it came with a multiwindowed, enlarged headpiece and robotic pincers. The miniature submarines featured in *For Your Eyes Only* were a mix of inventive props and the most advanced submersibles then available. Peter Lamont based his prop of a two-man submarine on existing research vessels, but the Osel Mantis one-man microsubmersible was the real thing. It was built in 1978 by Graham Hawks, an oceanographic engineer and designer who was responsible for the great majority of one-man submersibles used at the time. Described as a real life Q, and "the last of the gentlemen explorers," Hawks still builds futuristic submersibles and plans to take them to the depths of the oceans.[9]

Thunderball marked the point when Bond's equipment moved from the probable to the impossible—at least the impossible during the decade in which the films were made. The transistor and solid-state revolutions had reduced the size of electronic equipment, and the Bond films naturally reflected this. The homing device in his DB5 in *Goldfinger* is small enough to be fitted into the car's dashboard, and the transmitter can fit in the heel of his shoe (well before microchips would have made this possible), but by *Thunderball* it is a "harmless radio-active device" small enough to go into a pill that Bond has to swallow. The Geiger counter used by Bond in *Dr. No* is an off-the-shelf model the size of a toaster oven. By *Thunderball* the Geiger counters are small enough to go into an SLR camera. This piece of technological fantasy comes not from the designers at Eon Productions but from Fleming himself, who gives a detailed description of Leiter's "little Geiger toy," which appears to be a Rolleiflex camera, "but in the back of the make-believe there's a metal valve, a circuit, and batteries" (*TB* 133). It is connected to a watch whose "sweep hand is a meter that takes the radio-active count" (133). The filmmakers have gone much further than Bond's creator, for their Geiger counters are now small enough to fit into Bond's Breitling Top Time chronograph.

Thunderball marks a subtle but important shift in the uses of Bond's equipment. In the books he constantly has to improvise with whatever he has at hand, but in the films he usually has exactly the right equipment he needs, often with a brand name on it. In the book of *Thunderball* Bond and his colleagues fight the underwater battle with knives tied to broom handles, but in the film he has a multitude of specially designed weapons. Although Bond still improvises in the films, such as the old trick of throwing an electric appliance in a bath to electrocute its occupant, his equipment gradually becomes custom-made for the job, rather than thrown together in a moment of crisis. This moves Bond away from improvisation, the work of an inspired amateur, to that of a practitioner of advanced technology who employs devices ostensibly designed for the job by Q branch but really made up by the creative fabricators and artists in Eon Productions' prop department.

The most fantastic prop in *Thunderball* was the tiny rebreather apparatus used by Bond to swim underwater for long periods of time with only this small device in his mouth (thus evading Largo's sharks in his swimming pool). In this way he could carry out feats of derring-do without bulky scuba diving equipment on his back and was ready to make a sudden and dramatic dive underwater at any time without stopping to strap on air tanks and regulators. This device was made up in the special effects department by the simple expedient of gluing two small CO_2 canisters together. There was no chance that this device could work because there was no rebreather bag—Sean Connery and his stunt doubles just held their breath during the underwater shots. This gadget caught the attention of the film's audience, who marveled at the wonderful technology that enabled Bond the superhero to operate underwater as effortlessly as he did on land and in the air.

Life Imitating Art

Many of the Bond films' most avid viewers were members of the military and intelligence communities. When Ian Fleming was in Washington, he allegedly met President Kennedy at a dinner party and suggested that the Americans took Fidel Castro far too seriously and that the CIA should try and undermine his image through ridicule and innuendo. This argument reached the head of the CIA, and it is said that Allen Dulles tried in vain to reach Fleming to get a more detailed briefing. The CIA later developed these ideas into plans that included poisoning Castro with LSD and putting depilatory powder in his beard. The CIA's Technical Services Division came up with a variety of assassination devices to "get" Castro, devices that could easily fit into a Fleming plot: poisoned wet suits, exploding seashells, and a fountain pen charged with the Blackleaf 40 toxin.[10] Intelligence operatives all over the world enjoyed the Bond novels. The Egyptian secret service bought Fleming's books to use in its training, and when KGB agents arrested an MI6 officer and his secretary in Warsaw in 1967, a Bond book in her apartment was considered "a spy document." Bond as a vehicle for propaganda was quickly appreciated in the Soviet Union, and the KGB pushed for a Russian version to challenge Bond's global cultural preeminence.[11] The American intelligence communities were great admirers of Bond. Some sent fan mail; others made useful suggestions; and a few approached the films' producers with enquiries about obtaining some of Bond's equipment. Those at the sharp end of the business looked at each ingenious gadget with a professional interest, however improbable it might strike the viewing audience. The KGB defector Oleg Gordievsky revealed that "the Russians had seen the James Bond films. They were saying, 'Look at their technology . . .' I had lists and lists of the most sophisticated technological things the KGB was trying to steal." Robert Wallace, director of the CIA's Technical Service, noted that agents who watched the Bond films often asked: "Why don't we have all those neat toys and technical gadgets that Q makes for Bond?"[12] That agencies like the CIA and KGB actually experimented with devices they saw in the Bond films shows how influential these fictions were. Popular culture was still framing the technological development of spyware in the 1960s, as it had done in the 1920s and 1930s.

Bondmania made James Bond one of the icons of 1960s popular culture, and the modernity that defined the character made him a shorthand for devices so advanced that they challenged the viewer to decide if they were fact or fiction. "Just like a James Bond film" became a way of describing incredible, futuristic technology. The adjective *Bondian* had gained some currency within the intelligence community to describe a daring physical feat in the field. Looking back at his wartime activities in Spain, Kim Philby acknowledged that no one in SIS wanted the involvement of SOE sabotage teams there as they did not "welcome a James-Bond-like free-for-all in Spain." Philby had met Fleming during the war and thought of him as one of the "wild men" of special operations, describing him as "seething with wild passions." The kidnapping of the Nazi war criminal Adolf Eichmann by agents of the Israeli Mossad Secret Service in Buenos Aires in 1960 was called a "James Bond–style" operation. When BBC reporter Gavin Esler interviewed Rafi Eitan of the Mossad team in Tel Aviv, Esler suggested that he was Israel's James Bond: "He laughed, and said he was only half of James Bond." When Lt. Leonard LeSchack proposed Operation Coldfeet, his superiors asked him: "So you wanna be an American James Bond?"[13] As Bondmania grew, *Bondian* was applied as an adjective to high-tech equipment as his amazing gadgets conditioned people to expect previously unimagined feats of ingenuity from them. SIS operative Greville Wynne wrote after being arrested: "I suppose James Bond would have spat from his mouth a gas capsule (concealed in his molar), which would have overcome everyone but himself and would then have leapt to safety with a parachute concealed up his backside."[14]

The fact that professionals in the field were taken in by gadgets that were really props indicates how convincingly the Bond films showcased future technology. Surely, few in the audience really believed that anyone could be that irresistible to women or be able to escape from such impossible situations, but such was the technological enthusiasm of the 1960s that the most implausible gadget could be accepted as real as long as James Bond used it. After *Thunderball* Eon Productions had to field several inquiries about his rebreather gadget! Yet the amazing technological advances of the 1950s and 1960s had often made fact out of fiction, and this was Fleming's main line of defense against the critics who called his books, and Bond's gadgets, fantasies. He admitted, "My plots are fantastic, while often being based on truth. They go wildly beyond the probable, but not, I think, beyond the possible. Every now and then there will be a story in the newspapers that lifts a corner of the veil from Secret Service work." He then gives the examples of the Berlin tunnel that tapped Soviet military communications, Buster Crabb's frogman exploit, and the equipment handed over by the defecting Russian spy Nikoli Khorkov in 1953. This included several pieces of deadly Soviet spyware, including a miniature revolver that fired poisoned bullets and a cigarette case that fired dumdum bullets, similar to the assassination weapon Red Grant uses in *From Russia with Love.* Fleming concluded, "This is all true Secret Service history that is yet in the realm of fantasy, and James Bond's ventures into this realm are perfectly legitimate."[15]

The crash of guns, gangsters, attempted murder, were part of his job, his life.

The Spy Who Loved Me

012 Guns

Weaponry informs much of the Bond character because, as Ian Fleming tells us in *Goldfinger,* "It was part of his profession to kill people" (3). And Anthony Lane pointed out in the *New Yorker:* "Killing is his living, and his proof of life."[1] How Bond kills people and the equipment he uses help define his character: shooting his victim at a great distance with a sniper's rifle (which he does only once in the books) reflects differently on him than killing a man face-to-face with a pistol, knife, or bare hands. In *Doctor No* M says of the 00 section: "They've got to be properly equipped" (23), and the choice of firearms made by Fleming reflects what he admired in Bond and his own views on technology. The handgun is therefore the most important piece of Bond's equipment—so important that the silhouette of a long-barreled automatic pistol has become the primary image of the Bond brand. The first image we see of him in each of the films is through the barrel of a gun, and the person we see holds a pistol in his hand. The opening credits of *Dr. No* (created by Maurice Binder) formed a template for all future Bond films, fixing him in our minds and line of sight as the man with a gun. The cover art on some of the Bond novels was designed around a gun, and every film poster shows Bond with a pistol. A gun usually appears next to the *00* in film posters and art work, and in some of them the silhouette of the gun is added to the *007* to form a composite branding image. Bond does not point the gun or handle it in a threatening manner in these images but holds it quite insouciantly. Michael Hooks designed the poster for *Dr. No:* "I came up with that image of him with a gun. That came from that shot of Ian Fleming. . . . That was the inspiration for Bond holding a pistol in a kind of casual, debonair way."[2]

Ian Fleming was a gun enthusiast who liked to be photographed with them, but he was no expert. He consulted the gunsmith Robert Churchill to check that his descriptions of the firearms in the manuscript of *Casino Royale* were accurate, and Churchill corrected several mistakes, including the maker and the caliber of Bond's personal weapon. Fleming took some care in deciding what his hero should shoot, for Bond was, after all, "the best shot in the service," so he has to be properly equipped.[3]

Bond's Beretta and Walther were both employed by the enemy: the former by Italian troops, the latter by Germans. Fleming did not have his agent use the standard sidearm employed by the British army because its size and appearance did not fit the character he was trying to create. British officers had used the Webley pistol as far back as the Zulu Wars of the late nineteenth century. The Webley and Scott Mk VI was introduced in 1915. It is a big, heavy piece, firing the huge .455 round, which gives it quite a kick. It is too large to be concealed and certainly does not look glamorous. Major Smythe tells Bond in *Octopussy:* "Damned clumsy weapon. Hope they've got something more like the Luger or the heavy [large caliber] Beretta these days" (23). Fleming consequently chose a modern automatic pistol for Bond despite the fact that they are not as reliable as revolvers and thus not a good choice for a secret agent.

James Bond starts his career with the .25 Beretta automatic pistol—small, easy to conceal, and smooth in operation. Fleming notes that it is "very flat," which makes it difficult to spot when worn in a shoulder holster. (Bond always seems able to identify the bad guys by the bulges in their jackets.) The Beretta soon became a part of the Bond brand. The advertising copy written by the publisher Macmillan for *Diamonds Are Forever* ran: "Gentlemen may prefer blondes, but blondes prefer Bond, who is back with his trusty Beretta on a new assignment." But one reader was not impressed. A gun enthusiast called Geoffrey Boothroyd wrote to Fleming to complain about Bond's "rather deplorable taste in firearms," arguing that the Beretta was a "ladies gun" and the .25 bullet it fired did not have the penetrating power required by a 007 operative. Boothroyd favored a heavier revolver, the American-made Smith and Wesson Centennial model Airweight .38, "a real man stopper." Fleming responded that Bond was used to his Beretta, and he preferred accuracy over power, but added, "I am most anxious to see that he lives as long as possible and I shall be most grateful for further technical advice you might like me to pass on to him."[4] The next time Fleming was in New York, he visited the gun retailer Abercrombies and examined a Centennial Airweight, a small, snub-nosed revolver weighing less than a pound—comparable to the lightweight Beretta but firing a larger .38 Special round. Fleming could not get an export license to bring one to the United Kingdom. Strict gun laws and no gun lobby made handguns almost invisible in England, and not even the police carried them in the 1950s—only the very worst criminals and 00 agents. Fleming wanted a Beretta for the cover art for *From Russia with Love,* but could not locate one in England. Boothroyd finally lent him a Smith and Wesson .38, and this eventually brought a visit from detectives from Scotland Yard to check that the author had a valid certificate for his guns.

Fleming took the criticism of the Beretta seriously and got rid of "the bloody gun" in *Doctor No.* Boothroyd is called "the greatest small arms expert in the world," and he tells Bond: "Ladies gun, sir . . . No stopping power, sir, but its easy to operate. A bit fancy-looking too . . . appeals to the ladies" (*DN* 20). Boothroyd proposes the Smith and Wesson .38 for longer range work, and the Walther PPK for Bond's personal weapon. Boothroyd also sells Bond (and Fleming) on the Berns Martin Triple-draw holster, which has a spring inside to make it a quicker draw. Bond uses the Smith and Wesson to good effect in *Doctor No* when he takes on three gunmen: "Bond shot the rear man in the head and the second man in the stomach. The front man's gun was up . . . Bond's gun crashed. The man clutched at his neck" (*DN* 174). Bond can't resist a gibe at Boothroyd when the adventure is over, cabling "Smith & Wesson ineffective against flame-thrower"(*DN* 185), which refers to Dr. No's weaponized marsh buggy. Bond gave up the Beretta with some regret. He had used the gun for fifteen years, "never had a stoppage and I've never missed with it yet" (*DN* 22). Doing the math from the mid-1950s, when Fleming wrote the book, the Beretta was Bond's service handgun during the war. During "his fifteen years' marriage to the ugly bit of metal" (*DN* 22), it has certainly seen some wear, tear, and alterations. In *Live and Let Die* the Beretta is described as having a skeleton grip, but by the time of *Doctor No* Bond has wrapped tape around the handle to give a better grip and sawn off some of the barrel. It is his old reliable service pistol, just like the revolver Sherlock Holmes tells Watson to bring along on adventures of special danger. This worn, heavily modified gun is not only an important tool of the trade but also a repository of memories. Bond is justly reluctant to part with something that faithfully accompanied him during his war and the adventures that came after it: "How many times had it saved his life? How many death sentences had it signed?" (*DN* 22).

The Walther PPK became *the* Bond handgun and an important part of the Bond brand, appearing with him in the books and in all of the first 18 films except *Octopussy* and *Never Say Never Again.* Boothroyd lists its advantages: light trigger pull, the extension on the magazine makes it easier to grip, and more powerful 7.65 ammunition (roughly .32 caliber versus the Beretta's .25). Walther Arms Inc. of Thuringia in East Germany was a product of the Second Industrial Revolution, and its chief product was the automatic blowback design, which replaced the revolving mechanical cylinder in handguns at the turn of the century. Perhaps it was a coincidence that the Walther Model 1 .25 caliber pistol was introduced in the year of Fleming's birth, but there is a certain symmetry that the company's first automatic and its most important publicist appeared in the same year. The Model 1 was based on Browning's design and was intended as a personal-protection weapon small enough to be concealed. Walther scaled up its pistols for military use, and in 1915 the Prussian government ordered 250,000 Model 4s (7.65 mm, .32 caliber) for its troops. This was a popular choice for soldiers who wanted a smaller, more reliable handgun than the Parabellum Po8 (known popularly as the Luger), which was standard equipment in the German army at that time and for Bond's opponents ever since (despite its fragility and the difficulty in cocking it). Walther introduced the PP in 1929. It was designed for use by uniformed police (PP stands for Polizei Pistole) and was chambered for the 7.65 mm round in common use in police forces throughout Europe. A smaller, lighter model of the PP, the PPK (Polizei Pistole Kriminal) was intended for concealed use by detectives and other plain-clothes operatives. During World War II about 25,000 PPKs were issued to army officers, vehicle crews, and those in the security apparatus, in addition to the 45,000 that went into the hands of the police. In 1936 the PPK was adopted by the Nazi Party as the "Honor Weapon for Political Leaders."[5]

The PPK is a fine piece of German engineering, but arming Bond for the films was influenced by showmanship and the pressure to fit the weapon to the character. In the film *Dr. No,* when Bond has to hand over his trusty Beretta to M, it is not his small, ladylike .25 model 418 but the larger M1934—a more masculine-looking gun. Guns used as models for artwork and promotional material were chosen for their looks. One of the photo shoots for *Dr. No* produced an iconic picture of Connery with a long-barreled pistol held up against his cheek. This image was reprinted endlessly for merchandise and publicity, and no one challenged the authenticity of this prop, which was actually an air gun.[6]

James Bond's equipment is expected to move with the times, but nostalgia and tradition kept him using a gun designed in the 1920s well into the 1980s. In *Octopussy* he tells Q that he has mislaid his PPK and is given a 9 mm Walther model P5. This gun also appears in *Never Say Never Again,* which was also released in 1983. The P5 was much larger and more than a pound heavier than the PPK, and it sold very well for Walther, no doubt aided in part by the gun's association with James Bond. In *Tomorrow Never Dies* Bond uses the latest version of Walther's P series, which kept pace with other military/police sidearms with larger calibers, higher muzzle velocity, and a magazine that held sixteen

9 mm rounds, more than twice the number of Bond's original PPK. The P 99 is ergonomically designed and has a removable rear grip, which can be adjusted to fit the shooter's hand. Bond used it in three more films until he went back to his original PPK in *Quantum of Solace* and *Skyfall* in a nod to the Bondian tradition.

Secret agents use many different types of handgun in books and films, but all come with silencers—a necessary attachment for covert operations. Bond uses a silencer, and so do his enemies. The "Chigroe" assassins who kill Strangways and his assistant in *Doctor No* use guns with "sausage-shaped silencers" and "thick black silencers" (11). Fleming had plenty of experience with silencers during the war, when they were manufactured in SOE's workshops—noisy guns would quickly betray an agent in the field. SOE's gunsmiths continually experimented with systems of baffles and other mechanical devices to reduce the noise of a gun's discharge. A Polish officer, Lt. Kulikowski, developed the most effective silencer for the Sten gun, but the more it reduced the noise, the less muzzle velocity and penetrative power the bullet retained. Fleming remembered firing a Sten gun with a silencer, and "all one could hear was the click of the machinery."[7] But Boothroyd was dead set against silencers; they increased the length of the pistol, which made it harder to get out of a holster (the same objection that SOE agents had raised). He also told Fleming that as few people recognized the sound of a gunshot (this was well before television and films were full of them), it would not cause too much concern if it was fired in a well-insulated building. But silencers are essential equipment in spy films, no doubt because they change the shape of the weapon and enhance its phallic symbolism. And Bond uses them continually and to great effect. After he has shot Dr. Dent in the back in *Dr. No*, Bond laconically unscrews the silencer from his gun—a callous professional assassin at work.

When he took charge of SOE operations, Lt. Col. Gubbins started a program of designing special weapons for agents in the field. He had come up against IRA gunmen wielding Thompson submachine guns in southern Ireland and thought that this was the ideal weapon for guerilla warfare.[8] The Thompsons looked (and sounded) impressive, but they were too big and heavy for clandestine work, and their American .45 caliber ammunition was in short supply. SOE operatives needed a compact automatic weapon with a high rate of fire—a submachine gun—so the armorers at SOE's invention factories went to work. The Sten gun was inaccurate and unreliable, but it was cheap to manufacture and used the ubiquitous 9 mm round. It could be broken down into pieces, sent into the field via parachute, and reassembled easily—such was the simplicity of its design.

More than four million were manufactured, and it became the weapon most associated with SOE, commando, and resistance forces. It was still being used by terrorists and resisters in the 1970s. Although SOE's workshops came up with several improved submachine guns, such as the Norm gun and Welgun, they never challenged the supremacy of the Sten. With its notoriously light trigger pull, vulnerability to jamming, and tendency to discharge if dropped, it was not the ideal weapon for clandestine service, and perhaps that is why Bond never uses one. In fact, he never fires a machine gun or machine pistol in all his long career in the books; Fleming preferred his hero to make one expert, deadly shot rather than spray bullets around like his adversaries.

Fleming and Bond were similar to their World War II comrades in their preference for equipment of other armies: "Of course we had Schmeissers and Beretta pistols," said one SOE operative. Another agent noted, "We got the American carbine and a lot of American equipment, which we were very grateful to get because it was very much better equipment than ours."[9] Bond's backup handgun is a "long-barreled .45 Colt," which is hidden in his Aston Martin DB3 in *Goldfinger* and in the Bentley he drives in *Moonraker*, where it is described as a Colt Army Special. This weapon dates from the Colt "Peacemaker" six-round revolver, a fixture in cowboy films. Although it was replaced by the Colt M1911 automatic during World War I, demand for pistols was so high in World War II that Colt and other manufacturers produced them in some numbers, and many were rechambered from the usual .38 caliber to take the same .45 ammunition used in the M1911. They were distributed in the thousands to British and Canadian troops, and that is probably where Fleming saw them.[10]

Special Weapons

Single-shot assassination weapons figure large in the Bond books and films. During the war SOE had plans to eliminate some of the leading Nazis, including the Führer, and its workshops developed a series of one-shot pistols ideal for assassination. One of these was the Welrod, which came in 9 mm or .32 caliber versions with a silencer attached. Known as the "bicycle pump," it could be broken down into two pieces, which made it easier to conceal—a major consideration for this type of weapon. The Welwand was a single-shot pistol that could be hidden in the sleeve of a coat, hence its name: Sleeve Gun. The Welpen was another single-shot, "last resort" weapon. Known as the .22 caliber Experimental Firing Device, it was disguised to look like a fountain pen. The Welpipe and Welwoodbine were concealed in the stem of a pipe and cigarette respectively.[11] One can imagine Ian Fleming being drawn to these devices. With only one shot—it was impossible to eject a spent cartridge and insert another—you had to make it count, which suited Fleming's penchant for do-or-die situations. In *Casino Royale* one of the bad guys uses a one-shot, walking-stick gun: "Bond knew the type of gun. . . . They had been invented and used in the war for assassinations. Bond had tested them himself" (102) and surely Fleming had, too. This particular weapon fired a dumdum bullet, which has been cut across its face to accelerate its disintegration when it hits its target. Red Grant employs them in the assassination weapon with which he threatens Bond in *From Russia with Love,* and Valentin Zukovsky fires one in *The World Is Not Enough.*

Although James Bond earned his 00 status by killing a Japanese code expert and a Norwegian double agent, his missions rarely involve a planned assassination. He kills an assassin in *From a View to a Kill* and a former Nazi in *For Your Eyes Only.* In the latter Bond uses a Savage hunting rifle, "one of the new Savage 99Fs, Weatherby 6-by-62 'scope, five shot repeater with twenty rounds of high-velocity .250–3,000" (ammunition). Fleming describes how Bond gets the villain in the crosshairs of his scope and tells us that it isn't easy to keep the target acquired when sweeping the sight across the target area. The sniper rifle with telescopic sights came of age in World War II when the Russian and German armies employed regiments of snipers on the eastern front and some of their aces scored hundreds of kills. Fleming was intrigued by the technology of sighting through scopes, being particularly interested in the development of infrared devices to sight targets in low light. The rifle used by Kerim Bey to kill a Bulgarian agent in *From Russia with Love* comes in the form of a walking stick, with a "barrel from the new 88 Winchester," a silencer, and a "'sniperscope, German model . . . infra-red lens. Sees in the dark'" (128). In the film Bey uses an ArmaLite 7 rifle issued to Bond, which has a telescopic sight with night-vision capability. Designed by Eugene Stoner for a division of Fairchild Industries, the ArmaLite was one of the most advanced weapons available in the 1960s: a lightweight, highly accurate rifle that fired the compact, innovative .223 round. The AR-15 was supplied to the US military as the M16, and the AR-7 was a stripped-down version initially provided to aircrews for emergency use if they got shot down. The producers of the films always gravitated to the most modern weapons for Bond, while Fleming stuck with the tried and true; the cover of the novel featured a snub-nosed .38 revolver.

While Bond's métier is assassination justified by his 00 status, this is usually the role of the opposition, and most of the villains he confronts are out to kill him. SMERSH is defined as the "execution squad" of Soviet intelligence. In *From Russia with Love* Bond is cornered by the Soviet agent Grant, who has a gun hidden in a copy of *War and Peace*—"there are ten bullets in it—.25 dum-dum, fired by an electric battery. You must admit the Russians are wonderful chaps for dreaming these things up" (128). The throwing knives that came with Bond's special briefcase are out of reach, and as he lights his last cigarette, he thinks, "If only it had been a trick one—magnesium flare or something he could throw in the man's face! If only his service went in for these explosive toys!" (128). In real life it did. SOE produced a .22 caliber, one-shot gun concealed within a cigarette, an example of which was sent to the OSS, which also experimented with a similar weapon called the Stinger.

In Bond's world it is always the villains who carry out the aggressive covert action and killings, known as the "wet jobs," but the British and American secret services did their share of assassination in the 1950s and 1960s. During the Suez Crisis Prime Minister Eden demanded the assassination of Gamal Nasser: "I want him destroyed, can't you understand? I want him murdered." Stephen Dorril provides details of several ingenious plans to kill the Egyptian leader, including nerve gas, explosives hidden in Nasser's electric razor, poisoned chocolates devised by Q branch, and cigarettes that fired poisoned darts supplied by the chemical warfare experts at Porton Down. But as Anthony Verrier wrote, "the Prime Ministerial order was disobeyed. SIS was not living in a James Bond world," which shows how much it had infiltrated the world of espionage.[12]

Fleming's interest in scuba diving led him inevitably to harpoon guns; powered by compressed air, these weapons fire a barbed arrow with some velocity. They are intended to be used on fish, but Fleming and Bond could always be relied on to turn any device into a deadly weapon. A harpoon gun appears several times in the Bond stories. In "The Hildebrand Rarity" he has "a Champion harpoon gun with double rubbers" (299), and in *Thunderball* he is equipped with a couple of harpoon guns—"the French ones called Champion are the best" (64)—which he uses to dispatch one of the bad guys. His underwater equipment always includes a harpoon gun, as well as an underwater torch, commando dagger, and shark repellant.

Although he often carried a commando knife, edged weapons were not favored by Bond, but they were used by many of his opponents. What appealed to Fleming were the exotic poisons used on the tip of an edged weapon, and he describes a wide variety of them, including nerve toxins, developed by Nazi scientists, which the deadly Rosa Klebb smears on her knitting needles in *From Russia with Love*. Bond avoids this attack but Klebb finally gets him with the poison on the concealed knife blade in her shoe, which we learn later in *Doctor No* was fugu, taken from the Japanese globefish: "its terrible stuff and very quick" (18).

As an avid diver, Fleming was able to merge his fascination with poisons with his obsession with dangerous fish. They guard Mr. Big's treasure, maim Felix Leiter, and kill the unfortunate Major Smythe rather slowly—"fifteen minutes of hideous agony" (*OCT* 64). Ian was not the only Fleming intrigued by rare and exotic poisons. His older brother, Peter Fleming, was a writer, traveler, and intelligence officer during the war. His colleagues reported that he was "very keen on poisoned arrows."[13]

"Explosive Toys"

The deciding factors in Bond's fights are strength, ingenuity, and the right equipment. Facing Dr. No, "Bond had the usual blind faith that he would win the duel—all the way until the moment when the flame thrower had pointed at him. . . . Now he knew. This man was too strong, too well equipped" (130). After Bond kills one of the men sent by Dr. No to track him down, he picks up the man's rifle, a "U.S. Army Remington carbine, .300. These people certainly had the right equipment" (87). This rifle was hardly high-tech for 1958 as it was introduced back in 1903. Commonly called the Springfield, the .30 cal M1903 was a bolt-action rifle based on the successful Mauser design of the Second Industrial Revolution. During World War II, Remington redesigned it for mass production and produced the M1903A3 in great numbers. Thus, some American infantrymen were equipped with the same rifles their fathers had used in the Great War. The successor to the Springfield was the Garand M1, a far superior weapon with an eight-round magazine that automatically loaded rounds and thus fired much more rapidly. Nevertheless, Fleming preferred the earlier model—it was rugged, reliable, and accurate enough to be used as a sniper rifle. Michael Foot, who served in the Special Air Service during the war, approved: "It was simple, elegant and tough."[14] It was supplied to US Army Rangers and US Marine sniper units, and several of Fleming's commandos picked them up in North Africa and Sicily, where they served shoulder to shoulder with American troops. They were lighter than the English rifles, and their ammunition was plentiful. When escaping from Dr. No's compound on Crab Key, Bond chooses a Remington from a wide selection of weapons on a gun rack and uses it to wipe out the dogs on his trail.

Bond and Fleming show a preference for technology that is simple and effective, preferring the mechanical over the electronic. After Bond's car is disabled by steel spikes placed on the road by Le Chiffre's men in *Casino Royale,* "he reflected on the efficiency and the ingenuity of the equipment they used" (131). As Le Chiffre is beating Bond with "a three foot long carpet-beater in twisted cane," he remarks, "With this simple instrument, or with almost any other object, one can cause a man as much pain as is possible or necessary" (141). Emilio Largo is of the same opinion as he tortures Domino with ice cubes and a lit cigar: "These two simple instruments . . . applied scientifically" (*TB* 227). The evil scientist Kotze marvels at the "simplicity" of the atomic bomb Spectre has stolen in *Thunderball:* "even a child could handle these things" (105). Mr. Big in *Live and Let Die* has a mechanical lift under a trapdoor—an ingenious device employed in the very first crime and spy silent films, and one that turns up quite regularly in the Bond canon, where trapdoors lead to underground hiding places or shark tanks. Mr. Big has a gun fitted into a desk that he calls "my little mechanical toy" and "a sound technical achievement" (67). His practice of pouring offal in the sea to attract predators, who then guard his ship, brings Bond's admiration, who is surely speaking for Fleming when he describes it as "a typical invention—imaginative, technically foolproof, and very easy to operate" (185). These are the requirements of a field operative.

Although many products of SOE's invention factories were flights of fancy that never got much further than the drawing board, some of its greatest successes were models of simplicity. The staff of Station IX made an abrasive grease containing very finely ground carborundum, and when introduced into any machine with bearings, it was guaranteed to seize them up. A pair of French teenagers used this grease in early June of 1944 to disable many German rail tank-transporters. The four-pronged caltrop had a history going back to chariot races in the Roman Coliseum. Everett and Boyce called it "medieval,"[15] but it was a simple idea that rarely failed to burst a tire. In the novel *Casino Royale* Bond is chasing the villains in his supercharged Bentley when Le Chiffre's men operate a lever in their car that releases a steel-spiked, chain mail that forces Bond off the road: "he assumed it must be an adaption of the nail-studded devices used by the Resistance against German staff-cars" (131).

Bond likes to modify his equipment. He has his "pet expert" at headquarters install an Arnott supercharger in his Bentley, although he receives "solemn warnings" about blowing up the engine. He spends a lot of his time maintaining his Walther. In *Diamonds Are Forever* a long passage describes Bond as he "worked the action several times and tested the tension on the trigger spring as he squeezed and fired the gun" (488). He has made some modifications to it, filing down the firing pin and cutting the foresight from it. Fleming stresses that secret agents like Bond had to attend to the details of their profession, taking every precaution, like a "deep-sea diver or a test pilot, or to any man earning danger-money" (*CR* 16).

Bond is a master of improvisation who often deliberately breaks devices or works them to the breaking point—the way Fleming drove his cars. Much of the ingenuity in the books concerns the tricks and gadgets Bond uses to escape from imprisonment—for it is absolutely necessary to the plots of both books and films that he be in the hands of the bad guy at least once. Most, if not all, of the escapes have to be improvised. For example, in *Moonraker* he uses a cigarette lighter to escape from his restraints. Bond and Fleming have the ability to reconfigure everyday items into weapons or escape tools, such as the wire spear Bond constructs to escape Dr. No's torture course. Fleming delights in taking innocent machines, like the spine-stretching device, the Hercules Motorized Traction Table—in the health spa that Bond visits in *Goldfinger*—and turning them into instruments of torture.

Although Bond is not trained as a technician, he understands the workings of his equipment. He is able to drive any vehicle that comes his way, and he often finds a way to use any available machine to eliminate his enemies, such as his disposal of Dr. No with a crane; "Coolly, Bond reined the machine in, slowly easing the levers and pedals back" (*DN* 170–71). There seems to be no machine that Bond cannot operate, and his innate understanding of them is well illustrated when he realigns the gyroscopes in the Moonraker missile in a few seconds—a complicated task surely requiring expert and precise knowledge: "Press, click, and the tiny door had flicked open. . . . The gleaming handles beneath the staring compass-roses. Turn, Twist. Steady. That's for the roll. Now the pitch and yaw. Turn. Twist, ever so gently" (*MR* 425). Martin Willis concludes: "Central to Bond's power is his ability to master technology," and it is Bond's facility with machines that stands him apart from the spy heroes who preceded him.[16]

Nostalgia for more spacious, golden times
might be a source of revenue.

Casino Royale

013 The Special Relationship and the Cold War

Fleming's most lasting accomplishment of the war was his role in the formation of an American secret intelligence service. Britain's "special relationship" with the United States, a term coined by Winston Churchill in 1946, was to figure large in the career of James Bond and his creator. First and foremost 007 is a protector of the British Empire, fighting for what Fleming calls "Anglo-American civilization." Some commentators have argued that Fleming was essentially anti-American; he described the country as "Eldollaralado" and delighted in the tackiness of its lifestyle, yet he appreciated the grandeur of the continent's scenery and the modernity of its cities. His visits to North America and the Caribbean during the war provided him with experiences that he would weave into his Bond novels—his second and fourth adventures were set there. In late 1941 Fleming accompanied DNI Godfrey on a trip to liaise with the British Security Co-ordination (BSC) office in New York, the clandestine arm of British intelligence in North America. Under the leadership of William Stephenson, the BSC kept tabs on German agents, gathered intelligence, and provided material support for the British war effort. Most important, it worked to convince the American press, public, and politicians to join the fight against the Nazis at a time when isolationism was strong in the United States. In Stephenson's biography, *A Man Called Intrepid,* Fleming is described as participating in a Bondian scheme to break into the Japanese consular office to crack their safe and photograph their codebooks. He also allegedly took part in a plan to assassinate a British seaman who was selling information to the Nazis: "Killing him quickly perhaps saved hundreds of sailors' lives and precious supplies."[1] Violence was easily justified in defending Anglo-American civilization. Godfrey and Fleming also met with William Donovan and showed him the organizational plans for an American secret service. Fleming called Donovan "a close personal friend" and told

Donovan's biographer that his "memorandum to Bill on how to create an American secret service . . . [was] . . . in fact the cornerstone of the future OSS."[2] "Wild Bill" Donovan was convinced of the importance of secret intelligence and was like Fleming in his admiration for special operations, joining those voices close to the president who were advocating setting up units "for purposes similar to the British commandos and the Chinese Guerillas."[3]

The American intelligence effort reflected the British setup in that it divided special operations, which involved training and supporting resistance groups, from secret Intelligence, which was concerned with espionage. The special equipment was provided by the Scientific Research and Development branch. The man Donovan invited to run it was a chemist, businessman, and inventor with more than 70 patents. Stanley Platt Lovell was a member of the National Defense Research Committee, a group of scientists and engineers who had been charged by President Roosevelt to investigate new weapons technology. Donovan asked Lovell to invent "every subtle device and every underhanded trick to use against the Germans and Japanese," in essence to become the agency's Professor Moriarty.[4] The British sent examples of weapons, radio sets, sabotage gear, and other gadgets to Lovell. Henry Hyde of the Scientific Research branch maintained that "OSS owed everything to the British services. Everything—even such technical matters as suitcase wireless sets, one-time pad ciphers, and all manner of devices used by secret services came to us."[5]

OSS workshops produced the same submachine guns, special knives, secret inks, cameras, and timers as their British counterparts. They had their version of the explosive turd—explosive lumps of coal ("Black Joe")—and a few ideas of their own, such as an explosive flour called "Aunt Jemima." As befitting a chemist, Lovell commissioned experiments into female sex hormones to be placed in Hitler's food, chemicals that could put bloodhounds off the scent, and knockout pills to incapacitate enemy agents. Only the latter could be called a success, and they were manufactured in the thousands. OSS chemists and psychologists also experimented somewhat irresponsibly with truth drugs.[6] The British staffed their psychological warfare units with writers, journalists, and radio broadcasters, whereas the Americans employed psychologists and psychiatrists.

The start-up assistance provided by British intelligence was paid back many times over, with tons of equipment sent back across the Atlantic: radio transmitters, weapons, cameras, and a wide range of seagoing vessels, from kayaks to large submarine chasers. BSC even got hold of the exotic poison curare from Venezuela for some secret plan being cooked up in in London.[7] American expertise in electronics provided the vacuum tubes that were prized by English operatives, 12-line Hellschreiber teletype machines, and copies of Enigma machines that were manufactured by the National Cash Register Company of Dayton, Ohio. Lovell preferred to use the technical and manufacturing facilities of private companies to produce the items from original OSS designs or to modify existing products for clandestine use. While the British stayed with small-scale custom fabrication, the Americans used their mass-production expertise to set up long production runs. National Cash Register produced more Enigma replicas in six months than the total British output for the whole war.[8] By the end of the war, OSS had produced more than 25 pieces of special equipment. It published a catalogue of espionage and sabotage devices that could be ordered just like American housewives ordered crockery and furniture from Montgomery Ward or Sears. The Allied special equipment catalogues eventually ran to several hundred pages, and every item in this plethora of gadgets could be ordered by a product number. The Americans were justifiably proud of their gadgets, and Donovan sometimes took suitcases of them when he visited people he was trying to win over. He even offered suitcase radios and plastic explosives to the NKGB Russian security service in an abortive effort to build cooperation with them.[9]

The British shared their experience in training operatives and helped set up agent schools in Canada along SOE lines. American officers were sent to London to learn the ropes from SIS and SOE, enrolling in their training courses and occupying offices in their London HQs. This British education helped create a culture not unlike their own: founded on the old-boy network (this one based in Manhattan and the Ivy League), identified by a dress code, marked by excessive alcohol intake, and involving the same turf and class wars with the FBI that MI6 fought with the lower-class "tradesmen" of MI5. It was said that OSS stood for "Oh So Social." The British were happy to provide their expertise in return for the Americans' money and materiel. The latter were considered junior partners, inexperienced in the ways of espionage, as the writer and SIS officer Malcolm Muggeridge remembered: "arriving like *jeune filles en fleur* straight from a finishing school, all fresh and innocent to start work in our frowsty old intelligence brothel."[10] Some of them were impressed to be in the midst of the great British secret service; others found their opposite numbers arrogant and condescending. The OSS officer Joseph B. Smith described two schools of thought: one reflected the myth of the British secret service and its "long and remarkable tradition of success"; the other saw them as "a bunch of supercilious snobs."[11] There were the inevitable fights over turf. An agreement to keep out of European operations was soon broken as American agents and money were poured into the Continent. One SOE agent complained about "their permanent hankering after playing cowboys and red Indians," "their unlimited dollars, . . . [and] their capacity for blundering into delicate European situations about which they understand little."[12] Jealousy and distrust soon undermined the cooperation on which Churchill and Roosevelt had agreed. R. Harris Smith reported that many OSS officers operated on the general principle that "in intelligence the British are just as much the enemy as the Germans," and Muggeridge, who Philby said "had a good time sniping at OSS," bemoaned "how short a time the honeymoon lasted!"[13]

Fleming wrote the Bond books when the Americanization of European culture had become an issue, and he resented the erosion of English traditions by imported Americanisms. Cultural imperialism in the form of Coca-Cola and Disney was intrusive and hard to resist. Tiger Tanaka of Japanese intelligence tells Bond of the "Scuola of Coca-Cola: Baseball, amusement arcades, hot dogs, hideously large bosoms, neon lighting—these are part of the payment for defeat in battle" (*YLT* 80). The villain they oppose, Dr. Shatterhand, has castle HQ surrounded by his poisonous garden, which is described by Fleming as a "Disneyland of Death" (*YLT* 86). Americanization affected the culture of British intelligence in small but annoying ways. Replacing the stamps that said "Most Secret" with "Top Secret" to suit the Americans was one of many pinpricks to SIS's pride. The British often considered their allies naive in their reliance on technology. The British thought that evaluating new recruits was best done by taking them to lunch at the club, but "the Americans, on the other hand, so greatly concerned with 'psychological prognostications,' and often adopting pseudo-scientific 'evaluations,'" gave candidates a full battery of tests.[14] At the core of the resentment was the realization that the future of intelligence was in the hands of the Americans. Kim Philby spoke for many in The Firm when he concluded: "Whether the wartime exchange of British experience for American resources really paid off is matter open to argument. What is beyond doubt is that the decision in favour of co-operation doomed the British services in the long run, to junior status."[15]

Anglo-American Relations in the Cold War

The growing threat of Soviet Russia made it imperative to continue Anglo-American cooperation in the field of scientific intelligence. A few months after Churchill had made his historic "Iron Curtain Speech" at Fulton, Missouri, in 1946, Stalin made one of his own in which he predicted that war with the Americans and British was inevitable and that it only awaited Russian rearmament.

Allied intelligence knew that scores of Nazi scientists had disappeared behind the Iron Curtain and were desperate to know what they were up to. In 1948 an American committee led by Ferdinand Eberstadt came to the chilling conclusion that a "failure [to] properly . . . appraise the extent of scientific developments in enemy countries may have more immediate and catastrophic consequences than a failure in any other field."[16] The Allies continued their cooperation in aerial surveillance of the Soviet Union and in the fruitful exchange of SIGINT. The formation of the CIA in 1947 was followed by a formal UK-USA Security Agreement, which pledged cooperation between their code-breaking organizations.

The myth of British secret service took some heavy blows in the postwar years. Two scientists sent from Britain to the Los Alamos Manhattan Project, Klaus Fuchs and Alan Nunn May, were revealed as spies in 1950. Fuchs's arrest led authorities to a Soviet spy ring, which included Julius and Ethel Rosenberg, and started a "spy fever" in the United States. Worse was to come. The defection of consular official Donald Maclean and Guy Burgess in 1951 was a major blow to Allied intelligence as Maclean's position in the British embassy in Washington had given him access to atomic secrets. He also betrayed many of the operations that the CIA had mounted against the Soviets, making him one of the most effective Soviet spies of the Cold War. The identity of the "Third Man," who had warned Burgess and Maclean about their impending arrest, was hotly debated. Suspicion fell on Kim Philby, who ran the Soviet section of British counterintelligence and at the time of the defection of Maclean and Burgess was stationed in Washington as the liaison between MI6 and the CIA, which gave him access to the Americans' deepest and most cherished secrets. The majority view in London from men who knew and liked him was that Philby was a fine fellow, "one of us," and completely innocent: the majority view in Washington was that he was a Soviet agent and the person responsible for the most devastating breach of security in the history of espionage. Philby was now persona non grata in the United States. Under pressure from Walter Bedell Smith, director of the CIA, Menzies asked Philby to resign from SIS and hand over his passport. Bedell Smith had threatened to sever the intelligence cooperation between the two countries if Philby was not removed.[17]

Fleming incorporated these embarrassing intelligence failures into the novels: Fuchs and another Soviet agent, Bruno Pontecovo, are mentioned in *Thunderball*. In *From Russia with Love* he has Bond suggest that if "the atom age intellectual spy" (meaning Nunn May and Fuchs) were to be uncovered, the service must employ similar intellectuals who "understand the thought processes of a Burgess or a Maclean" and that their "closest friends" should be sent behind the

Iron Curtain to persuade the traitors to make full confessions and perhaps turn them into double agents working for the West (*FRL* 77). In a footnote Fleming points out that he wrote this in March 1956—well before some of Burgess and Maclean's closest friends were uncovered as Soviet agents, too. Fleming had started thinking seriously about writing a spy novel in the aftermath of the Burgess-Maclean defections and continued writing through the years of the search for the Third Man. The uncovering of SIS operative George Blake as a Soviet spy in 1961 came as another blow. He betrayed the CIA's Berlin tunnel tapping project, which had cost the Americans $25 million, scores of MI6 agents in Eastern Europe, and a CIA mole (P. S. Popov) inside the GRU (Soviet Army intelligence). Philby's escape to Moscow in 1963 ended the debate about his honesty. The CIA and FBI were appalled and blamed MI6 not only for harboring an operative who was obviously a traitor but also for allowing him to escape. The result was that MI6 access to CIA material was "sharply curtailed," which Stephen Dorril calls the "nadir of the special relationship."[18]

Fleming was writing *You Only Live Twice* when the Philby scandal erupted and probably learned the details from his close friend MI6 officer Nicholas Elliott, who was sent to Beirut to confront Philby with his treason in January 1963. Like many of his friends in the service, Fleming was appalled by his betrayal. He puts these words into Bond's mouth when he learns that an operative in SIS's Soviet section (which Philby led) is revealed to be a Soviet mole: "the covers which must have been blown over the years, the codes which the enemy must have broken, the secrets which must have leaked from the centre of the very section devoted to penetrating the Soviet Union" (*CR* 216). These lines are from the only Bond book to come close to capturing the deceit and betrayal that characterizes the secret war and the only mention of a Philby-like agent. Admittedly, every agent in the Bond books and films has a cover, but very few are revealed to be playing a double game. In Bond's world betrayal is rare, and to find an equivalent of the magnitude of Philby's treachery, Fleming would have had to have made Miss Moneypenny or Q a longtime Soviet spy.

Malcolm Muggeridge saw similarities between Philby and James Bond: "Philby, in other words, may be regarded as a real-life James Bond. His boozy amours, his tough postures, his intelligence expertise, are directly related to the same characteristics in Fleming's hero."[19] Fleming and Philby also had a lot in common: both worked in journalism and intelligence; both lived in the shadow of their fathers; and both liked to travel, entertain, and drink heavily. John Le Carré described Philby's "inbred upper-class arrogance, the taste for adventure," which were traits shared by Fleming. They had fond memories of the old, unreformed secret service of gentlemen's clubs and the old-boy network, described by Philby as "that happier Eden."[20] Yet it was exactly this old-boy system that Philby had so cleverly exploited. The culture of honor, hierarchy, and amateurism that Fleming valued so dearly proved to be the Achilles heel of his service.

James Bond and the imaginary service he works for were created by Fleming at a time when SIS integrity and reputation were shattered. In only two decades the intelligence relationship that Fleming had fostered had deteriorated alarmingly. He makes up for this disappointment in his Bond books, where he glosses over the reality of a "special relationship" ruined by the distrust of the Americans. Fleming preferred to relive his salad days of the 1940s. In *Casino Royale* M advises Bond that the "C.I.A. have got one or two good men at Fontainebleau with the joint intelligence [NATO] chaps there" (32). One of these is Felix Leiter, who becomes Bond's helper, his Tonto to Bond's Lone Ranger, in *Casino Royale, Live and Let Die* (where he loses some body parts in his support of Bond), *Doctor No, Goldfinger, Thunderball,* and *Diamonds Are Forever.* From the very beginning the relationship is made clear. Leiter tells Bond: "I'm under your orders and I'm to give you any help you ask for" (*CR* 62). When Bond loses all his service's money on the gaming tables in *Casino Royale,* Leiter comes to the rescue: "Marshall Aid. Thirty-two million francs. With the compliments of the U.S.A." (99). While the CIA was dismissing the discredited British security services, Bond is saving the Americans' space program in *Doctor No* and *You Only Live Twice* and their gold reserves in *Goldfinger.*

Nostalgia

The James Bond books reflect Fleming's nostalgia for a society and an intelligence service transformed by the war. A certain leisured but orderly way of life was fast disappearing in the 1950s, when Fleming started on the books. The British people had turned away from Winston Churchill to embrace the left-wing Labour government of Clement Atlee. For Fleming, moving to Jamaica was in a sense a retreat from a country in painful transition to an outpost of empire still relatively untouched by the postwar changes. His island retreat gave him, in the words of Major Smythe, a "glorious haven from the gloom and restrictions and Labor government of post war England" (*OCT* 45). Yet Fleming knew that he was living on borrowed time. Describing the Queen's Club in Jamaica, "the social Mecca of Kingston" in *Doctor No,* he laments: "Such stubborn retreats will not survive in modern Jamaica. One day Queen's Club will have its windows smashed and perhaps be burned to the ground" (3).

Ian Fleming the technological enthusiast made James Bond the perfect symbol for the emerging "New Britain": proud, resourceful, conversant with the latest technology, and unwilling to concede to the realities of postimperialism. Although broke and exhausted, the British faced the 1950s with a show of patriotism that climaxed with the Festival of Britain (1951), a celebration of British science and innovation. The world had entered the jet age, and no technology represented the future better than the sleek jet aircraft crossing the Atlantic in the 1950s. The De Havilland Comet, the world's first jet airliner, inaugurated commercial jet travel, which was heralded in the United Kingdom as a triumph of British inventiveness. With the country eagerly anticipating the coronation of Queen Elizabeth II, comparisons were made with the golden days of the Tudors. The chairman of the British airline flying the Comets proclaimed, "This present Elizabethan age is repeating in the air what the first Elizabethan era saw at sea." John Brabazon-Moore, now Lord Brabazon of Tara, said that the Comet reflected "the adventurous, pioneering spirit of our race. It has been there in the past, it is like that in the present, and I hope it will be in the future." Bond and Fleming flew the brand-new De Havilland Comets. Fleming tells us that "I rather enjoy flying" and was impressed with the Comets' performance, and the "varying whine of the jets" that replaced the deafening roar of the props,[21] yet Bond "did not care for it. It flew too high and too fast and there were too many passengers. He regretted the days of the old Stratocruiser" (*FYE* 226). The well-remembered days of luxurious, unhurried air travel were disappearing fast, along with such comforts as the sleeping berths in the double-decker Boeing Stratocruisers. In the autumn of his life Fleming the traditionalist was losing much of his technological enthusiasm.

The one part of Fleming's world that remained steadfastly traditionalist was "The Firm" for which he had once worked. The wartime expansion of the intelligence services, when the staff of SIS grew tenfold, had not changed the prewar culture of the organization very much. It still operated within the old-boy network and still valued the inspired amateur. George Blake said that "the pre-war secret service had been very much a kind of club of enthusiastic amateurs, autocratically ruled by the Chief." Nicholas Elliott was enlisted at roughly the same time as Fleming and rose steadily through the ranks of MI6. Like Fleming, he hated the paperwork, often disregarded the rules, and relied on "the British tradition of somehow muddling through despite the odds," admitting to John Le Carré: "we were terribly amateurish, in a way."[22]

Peter Wright was an electronics engineer recruited into MI5 after the war to work as the science officer in their technical services department. He described an intelligence service flush with their past successes but "covered in a thick layer of dust from World War 2." The Firm seemed quaint and archaic to outsiders; work often stopped so that personnel could adjourn to Lord's cricket ground. Wright had plenty of experience working with MI6, whose operatives he condemned for their "senseless bravado." He thought that they were "operating in the modern world with 1930s attitudes and 1930s personnel and equipment." Decades after the war, American visitors were still commenting on the homogeneity of class, attitude, and dress they found in the British secret service. Chester Cooper of the CIA noted identical black suits, Eton (school) ties, and tightly rolled umbrellas.[23]

Change had to come. Several committees recommended cutbacks and amalgamations of the intelligence services in 1945, leading to a period of painful upheaval. The financial attrition of the postwar years contributed "more than any other factor . . . to the amateurism of British Intelligence in the immediate post war era," according to Peter Wright.[24] Dramatic reductions in funding and bureaucratic reorganizations threatened the old guard, and every intelligence failure brought more pressure for change. In John Le Carré's *Call for the Dead,* his first spy novel, we are introduced to his central, abiding character George Smiley, a professor of seventeenth-century German literature. While a student at Oxford he was interviewed by one of his professors called Steed-Asprey and asked to join the "secret service." After serving in the Second World War in Sweden and Germany, Smiley is pensioned off, but the "revelations of a young Russian cipher clerk in Ottawa" necessitate his recall. The defection of Igor Gouzenko, a real-life confidential cipher clerk working for the GRU in the Soviet embassy in Ottawa, Canada, sent shock waves through the intelligence services in September of 1945. Not only did Gouzenko reveal Soviet spy rings in Canada and the United States; he also claimed that the highest reaches of British intelligence had been penetrated by Russian agents. Smiley sees that things have changed noticeably when he returns to the service: "Gone forever were the days of Steed-Asprey, when as like as not you took your orders over a glass of port in his rooms at Magdalene [College, Oxford], the inspired amateurism of highly qualified, under-paid men who had given way to the efficiency bureaucracy, and intrigues of a large government department. This was a new world for Smiley: the brilliantly lit corridors, the smart young men."[25] One of the smart young men was, of course, Kim Philby, who represented the new face of MI6's professionalism. He was the man of the future, destined to be "the star of the service." Yet there were still plenty of men from the past, and influential groups like the "Robber Barons" and "Three Musketeers" within SIS called for the aggressive covert operations of the old days and were not going to go quietly. Philby commented that the "haphazard and dangerously amateur" culture of the prewar service had been shaken up by the war, but it "took a long time dying. . . . SIS resembled the Chinese in their ability to absorb and digest alien influences."[26] SOE might have been closed down at war's end, but its derring-do values remained. Buster Crabb's foolhardy, unauthorized adventure was Bondian enough to fit easily into a Fleming story but without the usual happy ending. When Dick White was called in to clean house after this fiasco, one of his senior officers told him: "We're still cloak and dagger. Fisticuffs. Too many swashbuckling green thumbs thinking we're about to fight another Second World War."[27]

The bureaucratic infighting within the intelligence service, a major theme in John Le Carré's and Len Deighton's books, is significantly absent in Fleming's novels. This was not because Fleming was unaware of what was going on in the brightly lit corridors of power. Miss Moneypenny tells Bond that M is in a good mood because he has "won a bit of a victory at the FO [Foreign Office]" (*CR* 29). When Bond is told he is being given an assistant, "he hoped the man they sent would be loyal to him and neither stupid, nor worse still, ambitious" (*CR* 33). In *From a View to a Kill* Bond acknowledges that M worries about the independence of The Firm. The sudden removal of Fleming's boss came as a bolt from the blue and a timely reminder of the consequences of ignoring office politics. Godfrey was one of the men considered as a replacement for C, the chief of the service, but he had the unfortunate habit of speaking his mind, and this brought him into conflict with the prime minister, who hated to be corrected. This could have been the reason for his abrupt removal in 1942 and for "the astounding, not to say shameful, lack of any recognition of his immense services during the war, an omission which was, incidentally, deeply resented by every member of the Intelligence Division," as one of his subordinates concluded.[28]

Fleming must have resented this slight to a man he greatly respected, but he did not take the opportunity to retaliate in his books, as many other members of wartime intelligence did in their memoirs. In the James Bond stories there are no debilitating office politics, no stabs in the back from one's own side, and no relentless scheming of the office men to undermine the courageous work of the agents in the field. Instead, the men of action enjoy the fealty of the desk jockeys, and M rules the roost with hardly any opposition from above or below—a situation that may be the biggest fantasy of all in Bond's adventures. In writing them, Fleming recovered "that happier Eden" where men of integrity always triumphed. The values of Bond's world were personified by M, who was modeled on Godfrey. The DNI's driver was another old soldier from the war who comments after one act of consideration: "Just like the old boy. He'd always seen the men right first. . . . They didn't come like that any more" (*DN* 15). The secret service that Fleming knew and loved was staffed by the same upper-class characters from good families, who had gone to the best schools, joined the best clubs, and "counted life well lost for a thousand [pounds] a year," as Bond says in *Diamonds Are Forever* (537). Fleming knew men like these, cultivated their friendship as a useful source of information for his novels, and honored their values. As he was writing the books, he must have known that they didn't come like that anymore.

Although retired, Fleming stayed in close contact with old colleagues, joined in their gossip, and listened to their memories—some of which he reshaped in his novels. Fleming chose Istanbul as the exotic locale for *From Russia with Love* and incorporated the insider information that he got from Nicholas Elliott, who had been station chief there. The boozy reunions in London's clubs, where a lot of intelligence business had always been carried out, changed after the war. Where once the topics had been the current "game" against the Germans, it was now the cutbacks and retreats from expensive foreign operations. Getting the "chop" used to mean getting killed; now it meant removal from power and a severance packet. The last DNI before the position was abolished was Norman Denning. He remembered that during a reunion of intelligence operatives after the war "Ian said the whole organization wanted revolutionizing and he still saw a use for the traditional spy of fiction. Whereupon he imaginatively pictured the adventures he might have had in the last war given complete freedom and unlimited money. The exploits were not all that dissimilar to that of James Bond."[29] How wonderful it must have been to imagine a world in which all security services were united behind the British, "and of course money's no object. We can have everything, whenever we want it," as M tells Bond in *Thunderball* (80).

A vast program of modernization was due.

Peter Wright, *Spycatcher*

014 The Technological Revolution

Kim Philby began a talk to the members of the East German security service in 1981: "You have probably all heard stories that the SIS is an organization of mythical efficiency, a very dangerous thing indeed. . . . It honestly was not."[1] This was especially true of the technical support of operations. After the war there was general agreement that the equipment of the British intelligence services needed to be brought from the 1930s into the 1950s and their equipment and procedures modernized.[2] The technological advances of the war brought about important changes in the tools to acquire, store, and manipulate secret information, but Britain's intelligence services were laggards in adopting them. Peter Wright was horrified: "I realized then that MI5 officers, cocooned throughout the war in their hermetic buildings, had rarely experienced the thrill of a technical advance. . . . For the first time I began to realize just how bereft British intelligence was of technical expertise."[3] Wright's expertise was in the technologies of sound retrieval and storage—the microphones used in electronic eavesdropping and the tape recorders for saving hours of transcripts of interviews, meetings, and messages. MI5 used the equipment and employees of the British postal service, the GPO, in the clandestine recording of telephone intercepts. Messages were saved on Dictaphone cylinders—the same technology that Edison had introduced in the nineteenth century—and acetate gramophone discs, which dated back to the 1930s. When MI5 suspected foreign-office employee Donald Maclean of espionage in the 1950s, the agency put microphones in his house and taps on his phone. Teams of operators in a small booth in MI5 headquarters listened in and changed the acetate discs frequently as they recorded Maclean's conversations. Some recording was even done on the prewar technology of wire recorders, which had poor fidelity, were difficult to operate, and had a short recording time.

British intelligence clung to its outdated equipment and procedures. Record keeping might not be exciting enough to deserve a place in Bond's adventures, but it was being transformed by computing in the 1950s and 1960s. The SIS Central Registry contained all the information the service had gathered in years of espionage. The source books that recorded every agent and operation were bound volumes of paper—just like the reference books kept by Sherlock Holmes on Baker Street. Data was also recorded and cross-referenced on file cards. As the function of intelligence organizations is to gather information, they tend to accumulate large amounts of data. MI5's registry topped two million cards by 1940, and its "personal files" had jumped from 27,000 in 1918 to 170,000.[4] During the war the summaries coming from the code breakers at Bletchley Park were distributed into the files of each SIS subsection—comprising about five hundred thousand separate cards. Even a small organization like the BSC generated so many records during the war that it was estimated that it would take one man working eight hours a day, six days a week, sixteen and a half years to go through all of its registry. There was little enthusiasm for graduating to automatic data handling, using the new American Hollerith punched-card sorters. During the Blitz it was thought necessary to photograph every one of the Central Registry's file cards, which turned out to be a good idea as it suffered losses through bomb damage and several hurried moves. One of the obstacles to modernize data handling was lack of funds, because the collection of file cards was so large that any duplication of this mountain of information was bound to be extremely expensive. This was one of the reasons why there was "a holocaust of unwanted paper after the war."[5]

File cards had been the format of saving information since the creation of the secret services, and handling this archive required a good memory and long years of service to remember all the filing conventions and organization. The files were maintained and retrieved by a small army of upper-class women, the debutants who were always attracted to work in British intelligence. The communist George Blake noted that much of the SIS secretarial staff were "decidedly upper class and belonged to the higher strata of the establishment. . . . Though often scatterbrained, they worked hard because they were very conscious of their patriotic duty, instinctively equating the interests of England with those of their own class." Philby told a story about one secretary "who had a passion for cats and a highly personal filing system. When I asked her for a paper, she would say mildly: 'I think it is under the white cat,' and by God, it would be."[6] The operatives who used the registry every day remembered these women with affection and respect, and some agents, like John Le Carré, made them into characters in their spy fiction. "Connie Sachs," who plays an important part in *Tinker Tailor Soldier Spy*, Le Carré's dramatization of the Philby story, was reputedly based on Milicent Bagot of MI5's registry.

Computerization was the perfect technical answer to the problem of data overload, but replacing interesting characters with machines makes for poor drama; thus, it plays no part in the Bond narratives. Computers appear in the films only as part of the modernistic sets, not as a vital part of the equipment of modern espionage. Other spy fiction kept up with the times and its changing technology. Len Deighton's spy stories, for example, reflect the impact of computerization. In *The Ipcress File* he describes the acquisition of a new IBM card sorter by Major Dalby's section: "Dalby's latest toy—a low, grey IBM machine. . . . This IBM machine was the key to WOOC(P)'s reputation, for it enabled us to have files of information which no one could correlate except with the machine set the correct way. . . . It was quick, more efficient than humans, and it made Dalby one of the most powerful men in England."[7]

In addition to a culture that resisted change, the British intelligence service could not afford to make the transition to modern data collection and storage because of its cost. In contrast, their American "cousins" benefited from Cold War anxieties, which led to massive increases in funding. The CIA grew much larger and was joined in the secret war by the National Security Agency (NSA), the Defense Intelligence Agency, the Bureau of Intelligence and Research of the Department of State, the National Reconnaissance Office, and so on. Intelligence gathering became the great growth industry of the second half of the twentieth century. Nowadays, we talk about the US Intelligence Community, which covers sixteen organizations. The poor, beleaguered British could only watch in envy and frustration. The Americans had the technical expertise and the money to make use of the great leap forward in computer technology. The IBM 360 series made much more computing power available to businesses handling a lot of data, and the intelligence services were important customers. The Americans also took the lead in the technology of aerial reconnaissance. Whatever reservations they had about the shortcomings of their cousins in operations, the British had to admit that their equipment was the best: "One of the chief virtues of the C.I.A., in Bond's estimation, was the excellence of their equipment, and he had no false pride about borrowing from them" (*TB* 124). Malcolm Muggeridge would compliment the CIA/NSA operations that "came to outclass our once legendary Secret Service as a sleek Cadillac does an ancient hansom cab."[8]

The superiority of American equipment was partly the result of some important advances in electronics. The introduction of transistor technology in the

1950s addressed two of the most important issues in spyware: reliability and size. Replacing unreliable and bulky vacuum tubes with transistors led to new consumer goods like the transistor radio and tiny televisions but also to smaller homing devices and shortwave radios. Bond's equipment in the books comes with obsolete vacuum-tube technology. The "Homer" he places in Goldfinger's Rolls Royce had "a dry cell battery wired to a small vacuum tube" (*GF* 110–12), which were the weakest parts of the design and soon to be replaced with far superior transistor technology. The advance of solid-state circuitry led to even smaller electronic devices, which meant that they could be installed in items of everyday use, like clocks or telephones. This was a godsend to the creators of spy fiction and films, who were able to reduce the size of the equipment to such Lilliputian scale that it could fit into any prop at hand. The tiny Geiger counter display used by Leiter in *Thunderball* would have been impossible without the miniaturization of electronic circuits, nor would the "miniature buttonhole camera" used by the opposition to take photographs of Bond in *From Russia with Love.*

The CIA made developing new technology a priority. In 1951 it established a centralized research-and-development component to supplement the technical services organizations that provided the equipment for operations. In 1953 Dulles formed the CIA Research Board, with scientists and businessmen like Edwin Land of the Polaroid Corporation, to give the agency a perspective on the scientific research that might be useful in espionage. The CIA was to follow the OSS policy of exploiting the R&D capabilities of private enterprise in developing new equipment, but it had a much broader outlook in fostering scientific research over a much wider range of disciplines, including biology and psychology. The Doolittle Report on American intelligence capabilities in 1955 stressed that the United States should utilize "every possible scientific and technical avenue of approach to the intelligence problem . . . and destroy our enemies by more clever, sophisticated, and more effective methods."[9] When the CIA's Technical Services Staff was first established it had no more than 50 members, but when it was made into the Technical Service Division (TSD) in 1960, its staff was counted in hundreds and included psychologists and systems analysts, as well as chemists and physicists. The CIA took over the reconnaissance operation of the air force and eventually took it into space. Such was the importance and cost of the CIA's "Big Science" projects of aerial reconnaissance and SIGINT that a separate Directorate of Research was set up to deal with it. The growing distance between the Big Science of satellites and code breaking and the more mundane requirements of operations justified two separate R&D efforts.

In the United Kingdom the technical support organization was also divided up, reflecting the wartime division of scientists and craftsmen. The code-breaking effort was taken away from SIS and set up as the GCHQ (Government Communications HQ). Radio communications, SIGINT and electronic development went into GCHQ, while the invention factories of SOE and SIS were amalgamated into the Directorate of Research and Development. The remnants of SOE's Station XII continued to work on time-delay fuses, explosives, and secret inks just like they had done during the war. The storage and distribution of equipment was assigned to a separate office, and an army quartermaster was brought in to manage the latter and given the designation "Q." The Directorate of Research and Development was concerned with fairly mundane projects: better silencers, improved safecracking equipment, and "the destruction of paper"—all useful activities in the everyday business of espionage but with little of the ingenuity and imagination displayed by Fleming's Q.[10]

The CIA was full of technological enthusiasts and rightly proud of its advanced equipment, but along with its technological leadership came a measure of hubris. The men at Langley, where the CIA headquarters were located, were confident that they had the most advanced computers for cryptanalysis, the best spy plane in the world, and far superior aerial cameras. In contrast, the Russians were considered technological laggards; their clunky cars and refrigerators were objects of ridicule in the West. Fleming shared this view, for in *From Russia with Love* he has the directors of Soviet intelligence agreeing with the commonly held view that American equipment was the best. This was also the view of the KGB officers who defected to the West. Oleg Kalugin, a high-ranking officer from 1959 to 1989, admired the "featherweight" palm-sized radios used by the CIA and admitted that the bulk and weight of the Soviet equipment "were a great hindrance."[11] Although it had to be admitted that the Russians were good at setting up "honey traps" and blackmailing deviants, their technical capabilities were considered inferior. As far as the CIA was concerned, all they had to do was to maintain their aerial cameras and computers while they waited for another Soviet defector to walk into their arms and hand over the secrets.

This illusion came to an end when a USAF weather reconnaissance plane patrolling the Kamchatka Peninsula in the western Pacific in September 1949 picked up particles that revealed that the Soviets had exploded an atomic bomb in late August—16 months ahead of the schedule estimated by the CIA.[12] This was the first use of the science of radionuclide monitoring, which had been pioneered by Henry Lowenhaupt, a Yale chemist who suggested that evidence of a nuclear explosion could be evaluated this way. It joined seismic, hydroacoustic, and infrasound detection systems as the spyware of the future. By the mid-1950s the Russians had caught up with the Americans in jet propulsion, atomic power, and finally in aerospace, which was acknowledged as the high-water mark of technological expertise. The launch of their Sputnik satellite in 1957 was greeted with surprise in Washington, even though there had been plenty of advanced warning. Millions of Americans went into their backyards every night to get a glimpse of this light in the sky, which really began the space race. The resulting "Sputnik Panic" was largely the work of the media and politicians who took advantage of this loss of face, but it had great impact on policy makers and the intelligence community. Much like the improved strategic bombers of the 1930s, this advance represented a quantum leap in vulnerability and the threat of terror from the skies. The rocket that put Sputnik into space was an ICBM (intercontinental ballistic missile) that had a far greater range than the ballistic missiles developed from the V-2. This gave it the capability of reaching the United States and ended American security in the nuclear age in similar fashion that Bleriot's flight made the United Kingdom vulnerable from the air in 1909. All the fears that had driven English rearmament in the 1930s reappeared in the United States in the 1950s in the same dystopian form. Senator Lyndon Johnson warned that satellites would soon be "dropping bombs on us from space like kids dropping rocks onto cars from freeway overpasses," and one of his colleagues in the Senate announced, "What is at stake is nothing less than our survival."[13]

The Sputnik Panic led to an unprecedented investment in scientific and engineering education in the United States and to an acceleration of its own satellite program, which was directed at aerial photography rather than dropping bombs. President Eisenhower was speaking for all Americans when he said that nobody wanted another Pearl Harbor, and he funded a reconnaissance satellite program to replace the vulnerable U-2 spy planes photographing the Soviet Union. The first Corona satellite mission came in 1960, the same year that the Russians shot down Francis Gary Powers's U-2 and ended that program. This one Corona mission provided more usable photos than all the U-2 missions combined: 1,432 images, covering one million square miles of secretive and impenetrable Russia. Cameras beginning with the KH-1 were developed by the Land and Itek corporations, which provided images from 150 kilometers in space close to the resolution of the U-2 photos. The CIA and its customers in the military could only look in awe at the Corona photographs. Albert Whelan, the deputy director of science and technology at the CIA, commented: "It was as if an enormous floodlight had been turned on in a darkened warehouse."[14] The

Corona program changed the whole outlook of US intelligence, focusing it even more on Big Science.

Listening In

Russian science and engineering had also stolen the lead from the Americans in the technology of surveillance, which received a high priority in the closed Communist state. They moved away from the simple microphones that Fleming found in his Moscow hotel room in the 1930s to deploy directional microphones, which they used with some success during the Potsdam Conference in 1945. Until the 1960s, bugging was primarily "mic and wire jobs" in which microphones were hardwired to a listening station, making them easy to detect and forcing the listener to be as close to the target as possible.[15] These are the devices that Fleming himself encountered as early as the 1930s, and they appear in all the Bond novels, but the Russians soon advanced well beyond this. Their most advanced microphone, the surveillance equivalent of Sputnik, was employed in an American embassy, artfully concealed in the carved wood of the Great Seal of the United States, which stood right behind the ambassador's desk. The Thing, as it was called, looked like no other surveillance device; no wires or power source, just an inert metal and wood contraption. It confounded the CIA's technical staff until Peter Wright finally unlocked the secrets of what he named a "passive cavity resonator." The Thing's diaphragm top vibrated with sound waves, which created modulations on a radio signal being beamed onto the Great Seal and bounced back to the listening post, where the signal was deciphered and the sound duplicated. It was a brilliant piece of work that initially confounded Wright and Sir Frederick Brundett, the chief scientist of the War Office. Wright admitted that it represented technology that "was barely at the research stage in Britain." The Thing was a simple, elegant solution to the great problem of concealing listening devices. It was the work of the Russian inventor Lev Theremin—best known in the West for his theremin electronic musical instrument.[16]

Defectors from Russian intelligence proudly told their American interrogators about their advanced surveillance equipment, the "real high tech stuff we're developing." This included a clear liquid placed on top of vehicles, which could be seen from high observation posts; a powder called Metka, which they slipped into the pocket of diplomats to trace it on any letters or parcels they put into the mail; and a chemical called Neptune-80, which was placed onto the soles of shoes for specially trained dogs to track.[17] Soviet surveillance expertise was responsible for the arrest of Colonel Oleg Penkovsky of the GRU in 1962. Penkovsky had handed over to Western intelligence more than 100 rolls of exposed film, which contained secrets at the highest level of Soviet military intelligence. He gave intimate access to the Soviet arms buildup and proved that Nikita Khrushchev was bluffing when he announced that the USSR was turning out ICBMs "like sausages from an automated machine."[18] Penkovsky had provided information on Soviet intentions that could only come from an agent in the field, yet confirmation of the numbers of Russian ICBMs came from aerial reconnaissance—the U-2 spy planes flying high over the Soviet Union and photographing missile sites. Penkovsky's intelligence allowed the photographs to be correctly interpreted and guided the American response to the threat of Russian aggression during the Cuban Missile Crisis.

Penkovsky was an old-fashioned spy who used no technical equipment other than his Minox III camera and dead drops to deliver his film. The operation was run jointly by SIS and the CIA, and the British asset Greville Wynne and the SIS resident in Moscow were his contacts. It will never be known exactly how the KGB discovered Penkovsky's activities, but it is likely that trailing all foreign diplomats in Moscow led them to one of his dead drops. Once alerted to Penkovsky's spying, they bugged his phone and actually photographed him in his apartment training his Minox over a pile of documents. The handover of secrets from spy to handler is one of the most vulnerable parts of a spy's work, and many were captured as they charged or approached dead drops. Audio dead drops reduced this risk, enabling the operative to quickly transmit messages in bursts to a hidden receiver, which in turn transmitted it to a listening station. This was an example of new, "paperless" spyware, which worked so well that it "took off like gangbusters" within the CIA. The advances in electronic technology also made it possible to use the microphone as a transmitter, and in this way it could broadcast whatever it heard without the need for wires, which could easily be detected. The smaller size and increased capability of solid-state devices enabled engineers to design devices that scanned the airwaves searching for the transmissions of the surveillance teams.

In the film *Moonraker* Bond rifles through the handbag of CIA agent Holly Goodhead and finds what he terms "standard equipment": a pen with a poison needle, a diary that fires a dart, a perfume atomizer that shoots out flames, and a radio receiver with an antenna concealed in a handbag. This equipment comprised throwbacks to World War II technology and to previous Bond films. They are mechanical gadgets concealed in everyday devices, and the one electronic device can only receive rather than generate or detect radio signals. A few years previously a real CIA agent, Martha Peterson, was apprehended in Moscow by the KGB. What gave Peterson away was that she was found wired up to a handheld surveillance scanning device, the SSR-100, which was listening in to KGB surveillance frequencies. Peterson had been servicing a dead drop used by a CIA spy in the Russian Foreign Office, Aleksandra Ogorodnik, codenamed "Trigon." His CIA-issued equipment reflected traditional spyware: a poison pill in a fountain pen, a Zippo lighter with a tiny hidden compartment for film, a fake piece of concrete to use as a container in a dead drop, and a shortwave radio (bought off the shelf) to pick up messages. Yet he also had the newest, smallest, and most advanced camera the CIA had developed. The T-100 was small enough to be carried in a fountain pen, cigarette lighter, or key chain. A masterpiece of engineering, the T-100 finally caught up with the tiny cameras that silent films had shown spies using from the 1920s.[19]

Thus, while James Bond was still checking for hidden microphones behind picture frames in his hotel room, real agents in the field were using sophisticated electronic scanning devices to find them, while listening in on the bad guys who were following them. The bug attached to the wall and connected by wires to a listening station had been replaced by something much smaller, which needed no telltale wires. While Bond telephones his headquarters to submit important information, ground sensors in the field were now collecting and transmitting it silently and secretly. In Bond's world secret information is still sent by motorcycle courier, and the opposition has to go to the trouble of building an underground hideout with electric elevators to hide the agents who intercept the courier. In the real world of espionage, technical staff accessed the enemy's communication lines and slipped electronic eavesdropping devices onto them that broadcast the take back to HQ, where it was automatically recorded and filed by computers.

New Faces

The capture of Penkovsky was a major blow to the CIA, and it forced some drastic changes in its strategy and equipment. Clandestine operations behind the Iron Curtain were sharply curtailed, and the CIA embarked on a new program to replace vulnerable human agents with machines. One priority was the development of smaller cameras, which an agent could "use to photograph documents while inside a KGB rezidentura [headquarters]" without detection. The CIA benefited from an abundance of new technologies, such as the miraculous T-100 camera, and an injection of new blood. The CIA culture was changing from the "Oh So Social" of the OSS to a less elitist and more professional staff. Instead of Wall Street lawyers and Ivy Leaguers, new recruits came from state colleges and engineering schools and often knew more about science than their instructors at Langley.[20] New electronic and aerospace technology shaped Cold War intelligence and transformed the agencies that carried it out. The technical staff went from mere helpers, "the backroom boys" of the war, to the people whose equipment decided the success or failure of missions. There was considerable technophobia in all intelligence agencies during the postwar decades, but the technical staff finally emerged victorious. The rise of audio surveillance made them indispensable, and they were now at the top of the CIA's hierarchy. The field operatives who had once looked down their noses at their technical helpers were now deployed to look after the techies as they did their vital work, aggravating the tension between machine-derived intelligence and that of human agency. Willard Machle, MD, the director of OSI, spoke for the technical staff when he said he was "outraged that scientific intelligence should be dependent on clandestine collection by ignorant spooks."[21] Reliance on surveillance technology limited the opportunities for daring but often disastrous Bondian feats. The balance of power was shifting from the men of action to the desk and lab workers.

Fleming was well aware of the technological changes that were making his action heroes redundant. In 1960 he reviewed the film *Our Man in Havana,* which was based on Graham Greene's book, and pointed out how much the business of spying had changed while spy fiction remained the same. By 1960 "the spy is a ticking seismograph on top of the Jungfrau [mountain] measuring distant atomic explosions on the other side of the world, or instruments carried in aircraft that measure the uranium or plutonium contents of the atmosphere."[22] Luckily, Fleming did not live long enough to see the Bondian heroes he admired get relegated to backup and technical-support duties. He would not have recognized Allied intelligence services in the 1970s and 1980s: so much had changed. Penkovsky was serviced not by a tough-guy secret agent but by an untrained businessman (Wynne) and the wife of the SIS station chief in Moscow. Janet Chisholm took her babies to a park and waited for Penkovsky to come along and slip her a box of sweets, which contained his exposed film. The Penkovsky operation was decidedly low-tech. After Wynne was released from a Soviet jail, he ruefully commented: "I regret to reveal that the British Intelligence Service lags behind Bond in ingenuity."[23]

The End of an Era

The old ways Fleming revered were coming under threat from the modernization of espionage equipment and the resulting emphasis on technology rather than Bondian derring-do. The experience of the war had shown that SIGINT, code breaking, and aerial reconnaissance were the espionage tools of the future, especially as the Iron Curtain made the Soviet Union and its satellites difficult to penetrate with agents. SIS achieved a major success with signals intelligence when an underground tap of Soviet telephone lines in Vienna yielded much useful information. Working with experts from the GPO Technical Department at Dollis Hill, Peter Lunn (the SIS station head) had a tunnel dug and a tap connected to the underground cables that ran to the Soviet headquarters. The haul of material was so great that two more tunnels were dug, and more cable taps were planned for other cities. A 600-foot-long tunnel was secretly dug under the streets of Berlin, which gave GPO engineers access to 295 Soviet lines and a treasure trove of secrets. A bargain at only $25 million, this operation provided enough data to require 600 tape recorders operating continuously, fed by 800 reels of tape a day. The take was passed on to a grateful CIA. These successes led to a dream of "clean" operations without the uncertainty of operating agents in the field. George Blake of Y section of SIS noted that "senior officers in MI6 . . . believed that the future of spying lay in the technical field and that in time the human element would become less and less important."[24]

In the aftermath of the First World War, old soldiers like Winston Churchill would lament the mechanization of their trade and the replacement of their heroes with machine minders and lever pullers. At the end of the Second World War, old intelligence operators like Ian Fleming would lament the replacement of their wartime heroes with quiet men who listened in to telephone calls or examined photographs taken by satellites. Vested interests and the opposition to change in the intelligence services were strong. George Blake remembered "many old hands who shook their heads at all this new-fangled bureaucratic paraphernalia," and Peter Wright claimed that the modernization of technical support was resisted "with the greatest determination."[25] Fleming made Bond a gentleman, and as US Secretary of State Henry Stimson pointed out, "Gentlemen do not read other people's mail," but this was the future of the spy business. The well-dressed man of action was being replaced by a technician, an employee of the post office, whose tools were the headset and tape recorder rather than a pistol and a sports car.

Bond's uplifting heroics were Fleming's response to the decline of his service, its culture, and his country. Bond's enemies always taunt him about the end of British hegemony: Blofeld dismisses England as "a sick nation by any standards" (*YLT* 233), and Hugo Drax sneers at a country "too weak to defend your colonies, toadying to America with your hats in your hands . . . hiding behind your bloody white cliffs while other people fight your battles" (*MR* 408). The Suez Crisis of 1956 came at the end of a dismal decade for the United Kingdom: rationing, sterling crises, budget shortfalls, austerity, and loss of colonies. The humiliation of withdrawing from Suez weighed heavily on Fleming, Churchill, and the generation that had won the war. After the bungled invasion of Egypt Prime Minister Anthony Eden went to Fleming's house in Jamaica to recuperate, and the irony is not lost on Simon Winder: "At the zenith of national incompetence, the architect of this incompetence stays at the very house in which the greatest reassurance and palliative, the Robin Hood of British imperialists' darkest hour, was created."[26] Bond embodies the nostalgia of Fleming for a time gone by and for the myth of the British secret service: "Personally I am sufficiently in love with the myth to write basically incredible stories with a straight face."[27]

007: A gun and a radio. It's not exactly Christmas, is it?

Q: Were you expecting an explosive pen? We don't really go in for that anymore.

Skyfall

015 Into the Future

Part of the tried-and-true formula of the Bond films is the latest technology: Bond is an agent of modernism as much as of the secret service, and the films have become "a sort of library for the new or potentially new," as Simon Winder puts it.[1] This emerged as early as *Goldfinger,* when Bond's Aston Martin DB5 was the star attraction at the New York World's Fair. Broccoli and Saltzman worked with several retired military men, especially Charles Rushon (credited as "Production Liaison"), to bring the most advanced military technologies into their films, such as the Bell rocket pack in *Thunderball* and the MB Associates' rocket guns showcased in *You Only Live Twice.* As the franchise picked up momentum in the early 1960s, the space race between the United States and the Soviet Union brought public attention to rocketry and satellites. Technological enthusiasts who had dreamed of space flight at the beginning of the century saw it become a reality. Churchill and Fleming did not live long enough to watch the fiction of Verne, Méliès, and Lang come true on live television as a man walked on the moon (Fleming died in 1964, Churchill in 1965), but they did see weapons of terror evolve into a technology widely acknowledged as the ultimate achievement of humankind.

The Eisenhower administration created the National Aeronautics and Space Administration (NASA) in 1958. Ostensibly a scientific organization committed to advancing the technology of aeronautics and aerospace, NASA depended on military support as much as Wernher von Braun needed the Wehrmacht's money to finance his V-2 research. The rocket that would take a man to the moon was initially designed to deliver a nuclear warhead on the Soviet Union. With dozens of NASA launches every year, public interest increased. Astronauts were now global celebrities, and their journeys into space were broadcast on television worldwide. NASA's first launch in December 1957 was considered such a newsworthy event that it attracted more than 100 journalists to Cape Canaveral.[2]

This brought the technology of space flight into every home. Fleming's house in Jamaica put him close to NASA's main launch site on Florida's Atlantic Coast, and he incorporated this imposing structure and the procedures of the launch into his Bond books. In *Doctor No* he brings the test firing much closer to home in the Turks Islands near the Bahamas, but in the film NASA launches its rockets from Cape Canaveral. As the publically funded space agency NASA followed a policy of transparency in the interest of publicizing its work (and getting more congressional funds). This meant that its aborted launches were public knowledge, and in Fleming's account Doctor No's radio beam was behind the failure of the American launches.

The space race occurred as the Bond film franchise picked up momentum in the mid-1960s, and images of rockets and satellites fitted perfectly with Eon Productions' strategy of epic stories, widescreen images, inflated budgets, and massive sets. For patriotic Britons like Churchill and Fleming the challenge of space took them back to their imperial past and reaffirmed those manly, adventurous qualities they believed were the strength of their national character. Fleming was rekindling that old "spirit of adventure" in his Bond novels, encouraging the next generation to "steam out and off across the world again."[3] Broccoli and Saltzman were always looking for something bigger and better, and there was no better place for James Bond to ply his trade than the infinite reaches of the universe. Eon Productions duly took 007 into space.

You Only Live Twice appeared in 1967 at the height of the space race, and the film's marketing showed Bond with his gun held at the same jaunty angle while holding an astronaut's helmet. The film opens with an impressive scene of an American satellite circling Earth. Called the Jupiter, it was in fact a model of the two-man Gemini satellite—but by the time the film was made, this NASA program was completed, so a new name was chosen. The film's editors added in documentary footage of real launches and control rooms—complete with its staff of identically dressed technical staff with "McDonnell" (the name of Gemini's American manufacturer) displayed on the back of their shirts. As to be expected, the Russian space control center is much less elaborate, appearing rather dark and dingy compared to the bright lights and white coats of NASA's rooms. Televised NASA launches had introduced viewers to the launch control center and to the technicians operating the controls—the geeky rocket scientist craning over television monitors and working the banks of switches and buttons in front of him. Eon Productions sent its creative staff to NASA facilities to ensure that the control centers of the Bond films looked authentic, but incorporating this high-tech equipment into the set did more than shape the look of the film; it also helped tell the story. Radio communication had enabled early films to disembody the action by explaining the plot in radio messages, but television brought disembodied sound and images into the action and allowed filmmakers to use TV screens to tell the story. *Spione* and *Frau im Mond* had used fake screens, but by the 1960s, television was real and everywhere. Millions of viewers watched technicians in NASA control centers on their home televisions, while the technocrats themselves were watching astronauts on their wall-mounted monitors. The villains in 1960s spy films make their demands to the United Nations via television, and a nation of television viewers accepted this fantasy. Bond's world is remotely controlled (to open doors and blow things up) and constantly monitored on television screens. Everywhere Bond goes in Tokyo in *You Only Live Twice,* a camera follows him. Audiences, along with the good guys and the bad, see his adventures in real time on small television monitors set in desk consoles and in the Toyota 2000 GT sports car he drives. This established an important precedent in the action-adventure film: the communication and video technology employed by secret agents also serves the audience, who share the same screens and see the same action.

The centerpiece of *You Only Live Twice* was the villain's headquarters, which incorporated a launchpad, control center, a fully operational helicopter port, and a working monorail. It was the largest set ever constructed in Europe, and it certainly represents the crowning achievement of Ken Adam. At a cost of $1 million, more than the total budget of *Dr. No,* the set dominated the last half of the film. Adam's biographer, Christopher Frayling, argued: "there was an 'Adam look' . . . About halfway through the Bond films, they were actually constructing the scripts around his sets. He is as much an author of those films as anybody."[4] His massive set provided the climax to *You Only Live Twice,* when Bond's ninja commandos stage a World War II–style fight with Blofeld's minions, leading to the traditional ending of the set going up in flames and explosions. Not for the first time were members of the cast and crew told to rush over to Pinewood Studios because "we're going to blow up the set." *You Only Live Twice* extended the meaning of *Bondian* to cover grandiose sets and massive budgets. In *The Spy Who Shagged Me,* one of the cleverest Bond parodies, there is this piece of dialogue:

> Number 2: Your new lair is up and running.
>
> Dr. Evil: Is it a hollowed out volcano, like I asked for?

Producers justified the $10 million–plus budget for *You Only Live Twice* based on returns of the previous Bond films, which had surpassed $110 million by that time and easily trumped the competition's host of Bond-like action films.

An autogiro named *Little Nellie* was the star gadget of *You Only Live Twice.* It was a real aircraft built by a real inventor, Ken Wallis. Wallis was born in 1916 to a family of technological enthusiasts—his father ran a cycle and motorcycle shop and built aircraft with his brother. Ken spent his youth tinkering in his father's workshop, building his first motorcycle at age 11. His interest in aviation was piqued in the 1930s when he saw a Mignet HM.14 Flying Flea, a single-seat light aircraft, and this started him on the construction of his own tiny aircraft. Wallis was not a trained engineer but followed other ingenious amateurs in preferring "the bloody obvious combined with common sense" over academic theory. His home workshop was described as crowded with "mini cameras, scale models of bomb-loading trolleys, model racing cars and bits and pieces salvaged from German wartime jet engines."[5] Wallis built a miniature WA-116 autogiro for recreational use and jumped at the chance of publicizing it in a James Bond film. He did all the stunt flying in the film and also advised the special effects team about the weapons carried by this fragile craft. Wing Commander Wallis had once worked in the RAF's tactical weapons research unit at Boscombe Down, so *Little Nellie*'s equipment was right up-to-date with heat-seeking air-to-air missiles, rocket launchers (which would have completely destabilized poor *Little Nellie* if they had been fired the way they were in the film), and aerial mines. Bond is equipped with a helmet with built-in radio communications and a cine camera—a step forward in the technological enhancement of a character's becoming more integrated with his machines with each successive film.

Little Nellie was the perfect Bondian gadget: technologically advanced, compact, visually compelling, and available for purchase. It even came disassembled in custom-made crocodile-leather trunks for the discerning technological enthusiast. Wallis hoped that the exposure in *You Only Live Twice* would attract buyers, and in 1970 the AirMark Corporation started to manufacture his autogiro. Priced at £3,000, this piece of equipment brought James Bond's world within reach, but the development of helicopters in the 1950s and 1960s hurt sales. Nevertheless, *Little Nellie* proved more than a match for helicopters in the film as it destroys four Spectre machines one by one.

The idea of vertical flight was as old as the prospect of flight itself, but it took World War II to accelerate the development of single-rotor helicopters into a viable alternative to fixed-wing aircraft. Larry Bell established an aircraft company in 1935 and was so impressed by models of vertical takeoff aircraft built by the amateur inventor Arthur Young that he financed Young's work. His full-size helicopter first flew in 1942. The Vought-Sikorsky VS 316, based on the designs of Igor Sikorsky, also started production in 1942 and was adopted by the Allied armies. After the war Bell Aviation was at the leading edge of aviation technology, developing jet-powered, as well as vertical-takeoff, craft. It moved to the forefront of helicopter manufacturing with the military model 30, which evolved into the civilian Bell 47 and became the most widely used helicopter in the world. Felix Leiter pilots a Bell 47J to search the ocean for the downed Vulcan bomber in *Thunderball,* and in *You Only Live Twice* Bond is pursued by several Bell 47G helicopters, which were made under license in Japan by Kawasaki. When the Bond films appeared in the early 1960s, helicopters were coming into more general use, and like Roger Thornton in *North by Northwest,* Bond often finds himself fleeing from an airborne foe who sweeps down on him. Helicopters appear regularly in the Bond film series, from the heavy-lifting twin rotor Kawasaki-Vertol 107, which uses a large magnet to take the bad guy's car off of Bond's tail in *You Only Live Twice,* to the next generation of turbine-driven helicopters, which appeared in the Bond films of the 1980s and 1990s. Bond, of course, can fly anything, but given a choice (rather than being tied to the control column of an aircraft in a steep dive), he usually goes in for one-man craft such as *Little Nellie* or the Acrostar microjet that Roger Moore flies in the thrilling opening sequence of *Octopussy*—another high-tech toy constructed by an independent inventor.

As the most gadget-heavy film to date, *You Only Live Twice* moved into fantasy. Bond has his required briefing from M in an outrageously spacious, wood-paneled room in a diesel submarine. Bond's helper, Tanaka, has his headquarters on his private underground train, which travels around the Tokyo subway system undetected. The exteriors of the world summit meeting place were shot at a futuristic DEW (Distant Early Warning) radar dome in the Arctic—but the interiors were the pure German expressionism of Ken Adam. "Bird 1," the Spectre spacecraft that devours satellites in space, was designed as a scaled-down version of the Atlas ICBM used in NASA's Mercury program. In the film it is able to use its retro-rockets to reenter Earth's atmosphere and land precisely and safely on its lauchpad—a feat of rocket engineering that was not accomplished until the SpaceX company managed it in 2016.

Marketing and Merchandise

You Only Live Twice acknowledged Japanese culture, tourist landmarks, and its beautiful women, but it also showcased the ascent of Japanese manufacturing and technology, especially in vehicles and consumer electronics. In the late 1960s Japanese compact cars, televisions, and cassette tapes were dominating the American market. Once derided as cheap imitations, they were now considered superior to homegrown consumer electronics, not just in function but also in design. The clean, uncluttered lines of understated machines with smooth metallic surfaces suggested a sleek modernism that Ken Adam reflected in the sets of the film. Sony gladly accepted the challenge of putting Bond in the forefront of consumer technology by supplying some advanced equipment it was developing in its R&D laboratories. Sony provided the minitelevision and the telecommunications system installed in Bond's Toyota. Among the equipment donated by Sony (not all of which was filmed) were video cameras, a voice-activated tape and video recorder, and a cordless phone. Sony had just introduced its first video recorder aimed at the home market, the CV-2000, and managed to squeeze one (it was as large as a suitcase) into the car. The Sony brand name can be seen clearly as Bond operates these gadgets. Real agents in the field also benefited from affordable Japanese consumer electronics like reliable and inexpensive miniature cameras and tape recorders. Especially useful were shortwave radios, which were the best way to pick up messages hidden in normal transmissions (like the BBC bulletins used by SOE to communicate with its agents in France). The Japanese led the way into smaller, more powerful, and versatile receivers with interior antennas. No longer forced to string up 20 feet of wire and hide bulky equipment, the modern spy has only to pick the right time and switch on a Sony 2001 portable radio with scan tuning and extended battery life. As one of Len Deighton's characters remarks: "It's a lovely toy, that little shortwave receiver."[6]

By the time of *You Only Live Twice* the gadgets were an essential part of Bond's equipment and accepted as part of the formula for the films. Sean Connery had planned to retire from the Bond series after *You Only Live Twice*, and the next film, *On Her Majesty's Secret Service* (1969), was promoted as "gadget free" and "James Bond without the gimmicks," but the film was not a success. The lack of gadgets and a new actor playing Bond alienated audiences, so the producers committed themselves once more to Sean Connery, high-tech gadgets, and films that promised to be larger than life. Thus they returned to space in *Diamonds Are Forever*, which started production just after the triumphant moon landing in 1969 and was right up-to-date with the technology of the space race. The film featured a moon buggy developed by Willard Whyte's Techtronics in one of the obligatory chase scenes—three months before television viewers all over the world watched one being used in the Apollo 15 mission. The buggy was a prop based on the actual vehicle developed for NASA by Boeing and Delco. The director, Guy Hamilton, told Ken Adam that this machine, probably the most advanced piece of transportation equipment on the planet (and in space), did not look too "exciting"—a measure of the burden that "Bondian" put on the films' producers. Hamilton thought that the film required something that looked "more grotesque." So Adam added a clear bubble dome, a rotating antenna, and robot arms. His design was turned over to Dean Jefferies, whose car-customizing business had already produced fantastic vehicles for films (the *Green Hornet*'s Black Beauty) and television (the Monkeemobile). His concept car fitted the aesthetic of the Bond films, but it looked close enough to the real thing to convince viewers that this gadget was current in space technology. Bond, who proves that he can drive not only every earth-bound vehicle but also all those in space, uses the buggy to escape from bad guys driving Honda all-terrain vehicles (ATVs). The Honda ATC-3 three-wheeler was just being marketed in the United States, and this widely discussed chase scene helped the ATV gain popularity as a recreational vehicle.

Diamonds Are Forever appeared in the same year that the US intelligence agencies put their Hexagon satellites into space in the next step forward in global surveillance. Blofeld's satellite in *Diamonds Are Forever* had a more malicious function—using a high-powered laser to destroy missiles and submarines. The public's fascination with the laser used in *Goldfinger* had not gone unnoticed by Broccoli and Saltzman, and they used them in several other films. They claimed that *Goldfinger* introduced "a scientific device so new that only a minority of the general public have even heard of it," but their emphasis on style over substance greatly magnified the power of lasers, which could only cut through a razor blade at the time, as Professor Barry Parker points out. Scientists joked that the cutting power of the beam should be measured in "Gillettes."[7] The diamonds at the heart of the plot of *Diamonds Are Forever* are supposed to be used in a laser, but up to that time only rubies had been used. In *The Man with the Golden Gun* (1974) the villain, Francisco Scaramanga, has a solar-powered laser cannon, which comes with the "Solex Agitator energy converter," another invention of the filmmakers. By the time of *Moonraker* in 1979, the laser had clearly taken the place of the ray gun. Q now supplies agents with laser rifles, and the American marines who assault Drax's space station come equipped with shoulder-mounted laser rifles and handheld laser pistols. In *Licence to Kill* (1989) a laser is fired from a Polaroid camera—another example of how Bond's spyware becomes smaller, more concealable, and more fantastic with each film. Lasers became commonplace in Bond's world: mounted on vehicles to cut through metal doors (*Thunderball*); installed in satellites to destroy missiles, submarines, airplanes, and minefields (*Diamonds Are Forever, The Man with the Golden Gun, Die Another Day*);

or fitted in Bond's Aston Martin to cut through the bottom of a pursuing car (*The Living Daylights*). Bond even has a handy laser in his watch in *Never Say Never Again.*

In the third Bond space adventure, *Moonraker* (1979), the producers were able (with a little help from their friends) to shift the visual focus of their film from the rockets of the Cold War to the space shuttle, the reusable vehicle with cargo capabilities that was intended to go beyond scientific exploration to commercial space travel. Broccoli and Saltzman developed their space shuttle props, interiors, and models from NASA's plans and in time with the actual production schedule of the real vehicle—the idea was that the first shuttle flight would coincide with the premiere of the film. The shuttle was delivered by North American Aviation only three months before *Moonraker* premiered, but there was a delay in its flight testing, so the Eon Productions shuttle "flew" in a Bond film a good two years before its first operational flight.

Science Fact versus Spy Fiction

The author Roald Dahl, who worked as a screenwriter on *You Only Live Twice,* summed up the dilemma facing the film's producers: There is a "ready-made hero" on hand who needs little attention, but "the hard thing was to get a creditable plot without going into science fiction."[8] The audience believed in Bond as a technological maestro, and this gave the films license to push the envelope of reality. Bond's equipment had to come across in the films as believable, of being just a little ahead of the latest technology, but as early as *You Only Live Twice* Connery was complaining that these gimmicks had become so prominent that they were suffocating the character. The producers of the Bond films argued that they were representing "science fact, not science fiction," but in the 1970s and 1980s they gradually moved the franchise into parody and fantasy, a development most evident in the gadgets Bond uses. Thus, in the same film that Bond flies in a space shuttle, he drives an implausibly equipped gondola around the canals of Venice. Instead of one of the famous boatmen propelling it, Bond's gondola has a powerful inboard engine. At the touch of a button Bond engages a propeller under the craft that speeds it along the water. When he gets to St. Mark's Square, he presses another button, and the gondola grows skirts and becomes a hovercraft able to effortlessly cruise through the square.

Adhering to a formula brings continuity but also the risk of stagnation. *The Spy Who Loved Me* was the tenth film in the Bond series, and there was a sense of déjà vu in the repetition of plot, character, and gadgets. Ken Adam commented that he was "running out of ideas" for the design of the operations rooms and villain's lairs, while the set designers and prop men had exhausted their stock of gadgets. The production designers were now open to any idea for a new device, and any member of the crew who came up with a promising suggestion was accommodated, and this pushed the gadgets farther away from the possible or even the plausible. Martin Willis points out that Sean Connery's Bond depended on spyware that is often crucial to the plot and taken seriously by both characters and filmmakers, but "by contrast, Roger Moore's gadgets are played to melodramatic and slapstick excess."[9] Janet Woollacott reports that Christopher Wood, one of the screenwriters of *The Spy Who Loved Me,* asked his children what they liked about the previous films and then reworked the car chase that had thrilled them in *Goldfinger.* The chase scene involves the wonderful transformation of the Lotus Esprit from car to submarine, an impressive piece of automotive technology and filmmaking, but when Bond drives up the beach, he "casually and ostentatiously removes a small fish from the car, comically underlining [and undermining] the technological feat which he has just performed."[10]

When the films' producers enlarged the character of Q in the 1970s and 1980s, they chose to undermine his authority and make him more a comic figure than a vital part of Bond's work. Q's persona remains stuck in the past—the stereotyped "boffin" of the war—in attire and attitude. When he comments in *Thunderball* that it is "most irregular" to equip agents in the field, it had already become standard procedure. Sometimes his science is slightly off—for example, when he claims that it is possible to detect submarines underwater with heat-seeking location devices (*For Your Eyes Only*) and that the powerful magnet in Bond's Rolex watch can influence a speeding bullet (*Live and Let Die*), but as the bullet is made of lead this is currently impossible. The all-knowing Q of the 1960s gradually evolved into an often confused curmudgeon in the 1980s. His clothes are antiquated and somewhat tousled (compared to the studied calm and immaculate suits of his scientific opponents), and the producers delighted in dressing him in unflattering outfits, such as a Hawaiian shirt in *Thunderball.* The actor Desmond Llewelyn strongly objected to these outfits, especially the baggy shorts he had to wear, because he quite rightly pointed out that this form of dress was "completely against the character" and undermined Q's credibility as a master of technology. Over time Q leaves his rightful place in the Quartermaster's workshops and joins Bond in the field. In *Octopussy* he arrives on the scene in a hot-air balloon and wearing a World War I–era flying helmet, and in *Licence to Kill* he is reduced to playing the role of a gardener, with straw hat and false mustache operating a "broom radio," which makes him look more like Maxwell Smart than Merlin. By the time he operates a robotic remote-controlled dog in

The World Is Not Enough, it was clear that nobody was taking Bond's gadgets too seriously. When Desmond Llewelyn retired, he was replaced in *Die Another Day* by John Cleese, a comedian made famous by *Monty Python's Flying Circus.*

The astrophysicist Neil deGrasse Tyson noted: "What made the Bond gadgets attractive is that they used a little bit of what you already knew was out there, and just took it to some extreme, forcing you to say, 'Yeah, that could happen,' even if it had to violate a few laws of physics along the way."[11] We now live in a world powered in some small part by solar energy, which makes the technology featured in *The Man with the Golden Gun* in 1974 seem prescient. The filmmakers gave Scaramanga's Solex Agitator a 95 percent efficiency in turning the sun's rays into electricity, but Barry Parker points out that so far efficiencies of only 25–30 percent have been achieved.[12] Either by design or happy accident some of the gadgets we now consider commonplace appeared in a James Bond film well before they entered the consumer market. The electronic pager featured in *From Russia with Love* in 1963 had been invented in the late 1950s and had just been tested in small-scale networks operated by hospitals when the film appeared. Motorola's Pageboy I was the first consumer pager to be introduced. It appeared in 1974 and could only inform the user that a message had been sent. By 1980 there were more than three million pagers in use. In the same film Bond uses a car phone 10 years before Motorola's Martin Cooper made the first mobile telephone call from an automobile, and 20 years before they came into general use. In *Diamonds Are Forever* prescient biomedical devices are used to check identities. Tiffany Case inserts a photo of Bond's fingerprints into a scanner, where it appears in a compact viewer against an image of the fingerprints of the man Bond is impersonating. This should have been the end of Bond's disguise, but Q had supplied him with fake prints on false skins attached to the end of his fingers, a ruse anticipated in silent films. At this same time the FBI was developing the Indentimat system, which automatically checked for fingerprints drawn from criminal records. By the time of *For Your Eyes Only* in 1981, identification technology had been computerized in both intelligence operations and on film. Bond uses a digital imaging computer called the Visual Identigraph to automatically match facial features well before his real-life colleagues were able to take advantage of increased computer speeds to do so. In *Licence to Kill* Bond is issued a "signature gun"—which uses an optical palm reader to recognize his biomedical identity—decades before fingerprint locking systems in "intelligent" firearms were introduced. It was not until 2017 that a weaponized laser was tested by the US Navy.

The Seiko watches Bond wears from the late 1970s to mid-1980s represented the cutting edge of quartz-driven timepieces and were extensively modified in the films to act as spyware.[13] In *The Spy Who Loved Me* his Seiko 0674 works as a pager, and in *For Your Eyes Only* his Seiko Duo-Time H357 watch receives digital messages and can also transmit speech like a walkie-talkie. In *Octopussy* his Sports 100 G757 has a direction finder for a homing device, and his T001-5019 has a tiny liquid crystal television screen that displays images in color. Bond's watches demonstrate an increasing number of functions, which include working like pagers (with built in telex and teleprinter), radio transmitters, direction finders, homing device trackers, and video displays of surveillance cameras, as well as acting as a remote detonator for a detachable explosive device. All of this anticipated the functions of the smart phones and watches of the twenty-first century (even the destructive applications). The virtual-reality glasses that Bond and Moneypenny use in *Die Another Day* in 2002 seemed to be another piece of Bondian science fiction, but in 2016 Samsung introduced its smart-phone Gear VR, which allowed users to see their games and videos in a full 360° view.

Whereas the superheroes of the 1930s had supernatural powers, Bond has his gadgets. Equipment like *Little Nellie* and scuba-diving gear allows him to operate in environments—sky, space, and underwater—that are prohibitive to ordinary heroes. In this way he uses advanced technology to extend the possibilities of the human body, and like Batman and Superman he is able to effortlessly overcome the laws of gravity, physics, and human frailty. Some of the most memorable Bond gadgets, such as the Lotus Esprit automobile-submarine, have the capability to morph into something else. Although these miraculous transformations are the work of the films, Fleming had anticipated them in his fiction. Fleming wrote *Chitty Chitty Bang Bang* in the early 1960s, while he was convalescing from a heart attack. Chitty Chitty Bang Bang was an automobile modeled on a real racing car he had seen at Brooklands in the 1920s, but Fleming gave it the ability to fly, travel across water, and corner the bad guys.

Ian Fleming the technological enthusiast often imagined characters in his books as machines, and machines as living things. Chitty Chitty Bang Bang is a car with "a mind of her own," a vehicle that mends itself and comes up with "revolutionary and extraordinary adaptations" overnight: "She's got some ideas of her own."[14] When friends inquired about his health after his heart attack, he described the state of his body in these terms: "the mechanics (doctors) report that the engine, though less oiled than previously, is running on at least eleven out of its twelve cylinders."[15] Espionage organizations and their plans are described as machines: "Everything had been done that could be done, every precaution taken. The wonderful machine was running silently and full out." Fleming told his editor Michael Howard that "Smersh is, after all, a machine."[16] The novelist Umberto Eco has analyzed the plots of the Bond novels and identified a set of rules that govern Fleming's "narrative machine" with "the same precision and regularity each time." Eco argues that Bond acts like a machine and quotes Mathis the French agent from *Casino Royale:* "Surround yourself with human beings, my dear James . . . But don't let me down and become a human yourself. We would lose such a wonderful machine." Eco argues that Bond is unencumbered by "moral meditation" and never allows himself to be infected by doubt, concluding that Bond is "a magnificent machine, as the author and the public, as well as Mathis, wish."[17]

Patrick O'Donnell thinks that "as Bond becomes one with the car, or the mini-copter or the fashion accessories concealing knives and explosives, it is as if the totality of these resources and his expertise must be deployed in order for him to achieve totality as Bond."[18] In the twenty-first century the totality of the Bond character is achieved via biomedical engineering, as well as gadgets. Gene therapy and DNA transplants do the job that disguises and plastic surgery accomplished in earlier films, and they do so in a more spectacular manner, with props like the "Dream Machine" of *Die Another Day,* which flashes and pulses like the laboratory equipment of the old science fiction serials. Broccoli and Saltzman humor Fleming's obsession with exotic poisons in many of their films, but it is a robotic surgical system, the "Da Vinci" machine, and not a sexy nurse, who now repairs the wounded Bond. Colleen M. Tremonte and Linda Racioppi have argued that "technology becomes an extension of the gendered body, enhancing Bond's physical strength and prowess through the way in which he uses it."[19] The advance of biomechanical technology has extended Umberto Eco's point about Bond being a machine to encompass his physical being. The location devices that were once concealed in a cigarette lighter or the heel of his shoe are now implanted directly into Bond's body, as he and the gadget have become one in the twenty-first century.

> "You can find out all about a man, track him down . . . but you have to look him in the eye. All the tech you have can't help you with that."
>
> James Bond, *Spectre*

016 Keeping Up with the Times

The evolution of James Bond secret agent from his birth in Fleming's postwar angst to his long career and more recent incarnations in Broccoli and Saltzman's films reflect 50 years of spectacular change in technology, international politics, and popular culture. While the Bond books emerged from the specific historical and social context of the 1940s and 1950s, the films run from the Cold War scenarios of the 1960s up to the twenty-first century and beyond. Faced with a diminishing amount of James Bond fiction, Eon Productions moved away from Fleming's plots, and their films took on a life of their own, molded by current events and competing action-adventure films. While Eon's set designers and prop personnel took the gadgets from the probable to the impossible, their screenwriters took the social context and technological anxieties of the times and weaved them into the plots. They changed the MacGuffin of each Bond film to suit contemporary concerns about the latest technological threat. *A View to a Kill* (1985) reflected the rise of the integrated circuit and its growing influence on daily life in the mid-1980s. It was released in 1985, a year after Apple introduced its Mac personal computer. The plot involves cornering the market for microchips by creating a natural disaster in Silicon Valley, but whatever the threat, the consequences always remain the same: "He'll kill millions!"

The universe of bad guys created by Fleming from the binary oppositions of the Cold War was expanded in the films to embrace a variety of ills, some criminal, some political, and some just disillusioned with the state of the world. Karl Stromberg, the villain in *The Spy Who Loved Me* (1977), owes a lot to Captain Nemo of *20,000 Leagues under the Sea*. Stromberg is disgusted with the world in which he lives; civilization has become so corrupt and self-destructive that the holocaust he plans is easily justified: "I am merely speeding up the process" Like Nemo, Stromberg works in a futuristic underwater set, with luxurious

Victorian interiors and large windows that show the underwater world outside, but unlike Nemo—a man of action—Stromberg runs his operations by remote control from his command center. Working from a console on his desk (complete with a Sony television monitor), Stromberg pushes buttons to eliminate those who fail him in the ingenious but traditional ways we expect in a Bond movie—trapdoors that open into shark tanks, helicopters that are blown up by remote control, and the summoning of hulking bodyguards who have to be fought to the death. Yet these standard elements and equipment of the Bond story come encased in the contemporary framework of late 1970s economic ennui and concerns about the global environment. Stromberg's evil plan aims at destroying civilization above water while he builds a modern utopia beneath the surface, a new Atlantis—the name of his futuristic headquarters, which has the ability to operate above and below the water and also to provide striking, futuristic images.

A defender of the ruling order is defined by the enemies who threaten it. Cynthia Baron's examination of Bond's role in British foreign policy in the 1950s and 1960s concludes: "For Bond, the proof of his destiny as global policeman in a post-colonial world resides in the collection of villains he must face."[1] Fleming's villains usually represent the Other of crossbred neo-Nazis who are motivated by greed or revenge rather than ideology. The evil German scientists of the Fleming novels, the dangerous spawn of the war, continued to play supporting roles in the films. Dr. Metz in *Diamonds Are Forever* (1971) and Dr. Karl Mortner in *A View to a Kill* both have shady Nazi pasts, including medical experiments on concentration camp prisoners. But their orders now come from self-made millionaires and industrialists rather than ideological leaders or master criminals. Bond became one of the great heroes of 1960s film culture because he appeared at a time when the world stood on the edge of chaos and needed a savior. Throughout the 1960s Bond the secret agent / superman was always there to ward off the nightmare of an apocalyptic nuclear war and return things to the righteous status quo, but as Cold War tensions were relieved by arms limitation agreements (which were made enforceable by satellite surveillance) and détente in the 1970s and 1980s, Bond lost a lot of his raison d'être. As a locus of technological expertise, the multinational corporation filled in for the monolithic nation-states of the Cold War in the Bond films. The corporation had extensive technological resources, and the empires of crime visualized in the silent films of Lang and Feuillade were now placed into corporate hands. Fleming had already anticipated this shift in his novels, and Christopher Hitchens credits him as "a pioneer in moving [the villain] to crime cartels and non-state actions."[2]

The plot of *Diamonds Are Forever* revolves around the business empire of an eccentric American billionaire who was modeled on Howard Hughes. The producers were on good terms with the reclusive Hughes and were permitted to use some of his properties in the making of the film. The character of Willard Whyte has all the eccentricities of Hughes, but he does not suffer from paranoid delusions and is clearly on the side of the good guys. His private research organization, Techtronics, mirrored the many corporate entities providing the equipment and know-how for NASA. Hughes Aircraft was a leading contractor of the military industrial complex, manufacturing missiles and components of satellites and space vehicles. Drew Moniot credits a lot of the success of the Bond films in the 1960s to their antagonism toward the corporate state, a hierarchical, dehumanized organization wielding power through its control of technology and seemingly unlimited financial assets, and its indifference to the fate of its workers and customers, who are powerless to defend themselves against it.[3] The idea that a large corporation could be the agency of evil had emerged in the counterculture days of the 1960s, when corporate America was found to be selling dangerous products to the American people (as revealed in Ralph Nader's *Unsafe at Any Speed* [1965]) and profiting from the disastrous war in Vietnam. Respectable companies like Dow Chemical, who had sung the praises of "chemistry for better living," now had to admit to mass-producing napalm and the toxic Agent Orange.

Anxieties about the threat of noxious technologies added to growing concerns about the power of autonomous technological change. Writing in 1967, Robert L. Heilbroner tried to answer the question "Do machines make history?" and concluded that "what other political, social, and existential changes the age of the computer will also bring we do not know. What seems certain, however, is that the problem of technological determinism . . . will remain germane until there is forged a degree of public control over technology far greater than anything that now exists." Leo Marx used the term *technological pessimism* to describe the end of boundless optimism in progress and modernity. This came at the symposium "Technology and Pessimism," held at the University of Michigan in 1979, where scholars addressed the growing doubts about our ability to control technology, doubts that were undermining the basic articles of faith about the efficacy of technological change. They noted the decline in trust in the application of knowledge and in technological expertise and concluded that "technological pessimism has become an integral part of the emerging culture of postmodernism."[4]

While engineers confronted their own doubts about their role in technological change, strident criticism came from activists, environmentalists, and intellectuals about the threats of dangerous new technologies.[5] Public suspicion of corporations grew in the 1970s as the extent of their political power and dark secrets became better known. In *You Only Live Twice* (1967) the criminal organization is the Osato Chemical Company of Tokyo. One of the most appalling examples of industrial pollution had come to light in Japan, where the Chisso Corporation had polluted the waters around the town of Minamata with mercury, causing terrible birth defects to its inhabitants. Eugene Smith's chilling photographs of the victims were published worldwide. The lair of the villain in the Bond films of the 1970s and 1980s was more likely to be a conference room in a head office than a hidden location on a remote island or in an extinct volcano.

With the Cold War over and the Soviet threat receding into the background (except for a few renegade communists), the corporate villain became a staple in the Bond films. In *The Spy Who Loved Me* the bad guy is the industrialist Karl Stromberg; in *Moonraker* it is Hugo Drax of Drax Industries; in *A View to a Kill* it is the industrialist Max Zorin; and in *Tomorrow Never Dies* (1997) the media tycoon Elliot Carver is bent on world domination. Carver doubtlessly resonated with audiences, who were well aware of the global influence of media moguls such as Robert Maxwell and Rupert Murdoch. Corporate villains were sometimes replaced by drug kingpins, notably Franz Sanchez in *Licence to Kill* (1989), who is a bad guy with legitimate business interests. The Italian Mafiosi who intrigued Fleming and were popular in Bond films of the 1960s to the 1980s (*Goldfinger* [1964] and *Diamonds Are Forever*) gave way to Russian criminal syndicates and hackers in the 1990s, with *GoldenEye* (1995) and *The World Is Not Enough* (1999).

The new threat of global terrorism formed the backdrop of action-adventure films from the 1990s onward, and the Bond franchise followed suit. The opening scene of *Tomorrow Never Dies* is in "a terrorist arms bazaar on the Russian border." The psychopathic Renard in *The World Is Not Enough* is described as an anarchist, acts like a terrorist, and has previous links with the KGB; just as Fleming could not rid himself of World War II villains, the Bond film producers found it hard to tear themselves away from bad guys with bad Russian accents, and this gave character actors like Walter Gotell, who played the recurring character General Gogol, plenty of work. The chief villain of *Casino Royale* (2006) is a private banker for terrorists, and in *Skyfall* (2012) he is a cyberterrorist intent on revenge. The scientist standing behind the main villain is no longer a physicist or a chemist but a computer expert like Henry Gupta in *Tomorrow*

Never Dies, a radical American who we are told "practically invented techno-terrorism."

After *A View to a Kill* the aging Roger Moore stepped down as James Bond, and Eon Productions made two Bond films with Timothy Dalton as 007, *The Living Daylights* (1987) and *Licence to Kill* (1989), before a six-year hiatus—the longest in the Bond series—until they introduced a new Bond in *Goldeneye* (1995). During these eventful years communism had fallen in Europe, and there was little cold war or British Empire left to justify a 00 agent traveling around the world fighting "for England." Bond's world changed dramatically in the 1990s. There were new military technologies, such as stealth, to introduce and new weapons to arm the bad guys, such as deadly rocket-propelled grenades. Satellite surveillance had advanced to such a point that it was possible for an infrared image of Bond and his latest girlfriend to be transmitted from space in real time to the new, modern headquarters of British intelligence. There had also been some significant changes in the technology of everyday life: Bond's audience now had personal computers, mobile telephones, and a much wider range of consumer gadgets to entertain them. This new technological democracy began to influence Bond films during the tenure of Timothy Dalton and continued into the new century in Pierce Brosnan's tenure as Bond, which lasted from 1995 to 2002.

As James Bond approached the millennium, there was a return to a hero who took himself and his gadgets more seriously. Adventure films at the turn of the century, such as *Terminator 3: The Rise of the Machines* (2003), that depicted the threat of new, uncontrollable technologies reflected society's unease about mechanization, automation, and the unstoppable force of progress. While some people feared that advanced technology, especially artificial intelligence, posed a threat to humanity, others anticipated another type of threat in an apocalyptic breakdown of these systems that we now depend on. In his analysis of the Brosnan Bond, Martin Willis sees him as "a technological maestro who uses his virtuosic skills to alleviate increasingly hysterical millennial anxieties." Willis interprets the Bond films of the 1990s as a response to fears of the inadequacy of human agency to deal with increasing technological sophistication, reflecting the frustration and confusion of coming to terms with the digital revolution. As email, internet, and cell phones became an essential part of life in the 1990s, so did the headaches of operating them. In *Tomorrow Never Dies* Q fumbles with the remote-control device that operates Bond's BMW, and Pierce Brosnan takes it from him and works it with ease: "Bond's mastery of this piece of modern technology, humorously compared with Q as the baffled boffin, serves to put the audience at ease."[6] This scene was doubtless a reflection of similar moments in millions of households when children explained the workings of the latest electronic gadget to their bemused parents. The technological threats Brosnan's Bond faces reflect the high technology of the information age of the 1990s—cybernetics, digitization, and mobile computers—and the accompanying concern about the ability of technology to compete with human intelligence. The villain Elliot Carver plans to use the power of his media empire rather than nuclear weapons in his plan for worldwide domination, claiming, "Words are the weapons, satellites the new artillery."

Bond at the millennium still uses some of the equipment he inherited from SOE's quartermasters, but it is updated to function in the digital age and branded appropriately. Where the postwar Bond carried simple escape tools in his key ring, Timothy Dalton in *The Living Daylights* has one that contains a gas that stuns anyone standing close, a detonating device, and an apparatus that can open "90 percent of the world's locks." This piece of spyware has the "Philips" logo prominently displayed, as does his special car radio, which can scan police broadcasts. Bond uses a magnetic device incorporated into a credit card to open up a window in *A View to a Kill* that carries the name of "The Sharper Image," a business that sells high-tech consumer gadgets.

The Bond films of the 1990s still depicted the usual technological staples—space-based weaponry, nuclear weapons in the hands of renegade Soviet generals or terrorists, and digital MacGuffins in the form of red boxes and microchips—but Bond now operates in a world of smart phones, laptops, palm-sized computers, and internet connections. Rather than a treasure map or tape cassette, vital information is now digitally encoded in computer files or in the CD-ROM that holds Felix Leiter's secret files in *Licence to Kill*. The spy of the 1990s was more likely to use email or video links to communicate than secret ink or hidden wireless transmitters, and we see many more shots of Bond sitting at a computer during this decade. Bond's chief piece of equipment in *Tomorrow Never Dies* is an Ericsson JB988 mobile phone, which comes with a 20,000 volt stun gun, fingerprint scanner, and electronic lockpick. This phone has a touch pad and LCD screen, which Bond uses to drive his BMW 750 by remote control. Bond and his adversaries need new skills to operate their equipment—skills honed by playing computer games. Remote control allows Bond to use his vehicle as an assistant who can look after itself (with a shield of electricity) and knock down the bad guys with rockets and machine guns.

The rise of digital technology also affected the way audiences viewed Bond's adventures. Computer screens are more and more used to tell the story, define the characters, and explain the equipment. What the audience often sees in Bond films are screens in which characters watch other characters on screens. A critical action scene in *The World Is Not Enough,* when Bond and the girl disable a nuclear bomb in a pipeline, is seen partly in live-action photography and partly on screens in MI6's headquarters in the form of electronic displays.

While Eon Productions showed Bond negotiating the shift into digital technology with ease, the franchise fared far worse in dealing with the digital revolution in filmmaking. Computer-generated images (CGI) transformed the making and the look of action films. *Moonraker* was produced under the shadow of the immensely successful *Star Wars* (1977), which forced the Bond films to keep abreast of the technological developments in both space travel and motion picture special effects. Bond-fan Simon Winder dutifully sat through *The Spy Who Loved Me* but "jumped with delight at Star Wars. We were there at the birth of the modern, globalized Hollywood blockbuster and just loved it."[7] He remembers that by the late 1970s the Bond films had lost their aura as motion picture events, which the competition had taken to the next level with blockbusters like *Jaws* (1975) and *Star Wars.* Putting more sharks into Bond's adventures and naming one of the villains "Jaws" was not enough to compete with them. Despite showcasing high-tech gadgets, the Bond films began to look dated in the 1970s. The sex and violence quotient that came close to the edge of censorship in the early 1960s stayed constant thereafter to keep the PG rating essential to draw a mass audience, while the competition increased the amount of sex and violence that was permissible on film. Improved special effects made violence look much more realistic, with exploding blood packages and remote-controlled prosthetics, but the Bond franchise stayed quaintly old-fashioned with hardly any blood or gore; the villains duly fall over like kingpins, and Bond continues on his way. Eon Productions faced much stiffer competition from action-adventure films that displayed a lot more violence, romantic films that showed a lot more flesh, and science fiction films that raised the bar of special effects. Bond spent the 1980s just going through the motions. With lengthy establishing shots, languid editing, and the sound dubbed in during postproduction at Pinewood Studios, *A View to a Kill* looked more like a tepid 1960s Hollywood epic than a film that could stand up against Superman, Indiana Jones, or Robocop. After being the undisputed leaders in stunts, the franchise suffered when CGI leveled the playing field for effects-driven films because now anyone was capable of doing Bondian feats with the help of a little computer animation.

Boys' Toys

Kingsley Amis coined the phrase "the Fleming effect" in his argument that the science of Fleming's science fiction is in the details. Umberto Eco and Christopher Hitchens have pointed out that the technical details and use of branded products serve to take the readers' mind off what Hitchens called "the glaring lacunae of the plots." Aaron Jaffe sees 007 as "a form of fictionalized reality, real insofar as it is defined by detailed attention to brand-name commodities."[8] Ian Fleming was one of the first writers to specifically name products in his novels. A renowned name-dropper, he filled his plots with luxury goods, as well as spyware, arguing that "the use of branded names in my stories helps the verisimilitude, so long as the products are quality products." Broccoli and Saltzman agreed. They realized that Sean Connery had very little name recognition in the United States in 1962, so they made deals with up-market cigarette and liquor brands to build up his Bond persona. By 1965 there were 39 different companies with licenses to use the Bond brand, and they made everything from underwear to action puppets of Odd Job. In Sean Connery's words: "The whole thing has become a Frankenstein monster. The merchandising, the promotion, the pirating—they're thoroughly distasteful."[9]

Machines are often described as toys in the Bond novels. Blofeld describes the nuclear arms race in terms of "Rich boys playing with rich toys," and weapons of mass destruction are "dangerous toys" (*YLT* 232–33). In *Thunderball* Emilio Largo calls the atomic bombs he stole toys taken from the toyshop, and in *Moonraker* Fleming describes the ominous V-2 rocket as "beautiful, innocent, like a new toy for Cyclops" (400). Military men often call their equipment toys. Lord Kitchener called the first tanks "pretty mechanical toys." Michael Foot described some of the gadgets produced for SOE as "those over-ingenious toys that kept its devisors out of mischief, but did little else useful for the war effort."[10] This term was widely used in the intelligence services. Toys were equally important for Bond's business affairs because playthings for children formed the major part of 007 merchandising in the 1960s and 1970s. Bond's image appeared on board games and action puppets, but models of his vehicles and gadgets were the most popular. Bond's DB5 was reproduced in battery-operated models and in plastic kits, but by far the most successful version was made by Mettoy, a manufacturer of die-cast model cars. Its model of Bond's Aston Martin was highly detailed, came with working ejector seat, and was named "Best Boys Toy of 1965." Scores of DB5 models are listed on eBay, some selling for as much as $500.

By the time of *Thunderball* it was understood that any exposure in a Bond film brought your product worldwide attention. The Bond brand had already been applied to a wide range of merchandise, either reproducing an item from the films, such as Airfix's plastic scale model of *Little Nellie,* or trading in the Bond cache with fragrances and clothes, but a product placed directly into the films' action brought a much more powerful association with 007. After the runaway success of *Goldfinger* in marketing Aston Martins, Broccoli and Saltzman had little trouble in getting whatever cars, aircraft, boats, and submersibles they needed. The marketing manager for the British racing concern Lotus deliberately parked an Esprit S1 in the parking lot of Eon's offices and waited for someone to take the bait. Lotus got the call and duly provided two Esprits for use in *The Spy Who Loved Me* plus five bodies for customization—a bargain in return for the Bond imprimatur. Manufacturers were willing to go one further to supply Bond with special customized equipment. Toyota built a soft-top version of their 2000 GT sports car specially for *You Only Live Twice,* and from then onward, the status of providing Bond's vehicle came with a commitment to work with the producers to customize it to Q's requirements.

As the Bond franchise reached new heights of popularity and new foreign audiences in the 1970s and 1980s, its value to advertisers grew, and product placement began to play a much larger role in the films. Initially, the franchise featured only expensive playthings for the very rich: corporate jets, flashy speedboats, helicopters, and watches too expensive for the average cinemagoer, who can only sit in the theater and yearn for these luxury toys. In *Goldfinger* Pussy Galore's Flying Circus takes to the air in Cessna 140s, an affordable prop plane favored by amateur pilots, but she flies Mr. Goldfinger around in a Lockheed Jetstar—a twin-engine corporate jet that had been introduced only a few years before production of the film started. This was the most luxurious private jet on the market, with a top speed of Mach .8 and a range of 2,000 miles. *Thunderball* introduced viewers to the wonders of underwater cameras, which had been developed by the Japanese camera-maker Nikon in cooperation with Jacques Cousteau. The Nikon Nikonos I was a custom-built camera that Cousteau popularized as the Calypso-Photo. By 1963 Nikon had developed this advanced camera into a commercial product for underwater enthusiasts, just two years before it appeared in *Thunderball.* In the film the camera can take photos underwater with infrared film (but with no flash), which underwater cameramen know will not produce usable images, but this didn't harm sales of Nikonos cameras.

No commercial partner fitted the Bond brand better than *Playboy* magazine, which considered Fleming's hero "the ultimate material man." *Playboy* had immediately jumped onto the Bondmania bandwagon, serializing several of Fleming's stories and featuring Bond's gadgets, girlfriends, food, drink, clothes, and vehicles in its pages. Clair Hines has catalogued their symbiotic relationship in promoting hedonism, international travel, and technological gadgets. *Playboy*'s readership looked for information about conspicuous consumption for discerning masculine tastes, and Ian Fleming duly provided this with his own experiences of living the good life.[11]

Style is at the center of Bond's appeal to advertisers, as a Finlandia vodka advert explains: "James Bond is only associated with the best things in life: the best cars, the best women, and the best vodka." Yet Bond's style, as revealed by his dress and savoir faire, does a lot more than establish his place in the social and cultural hierarchy; it has become a factor in the success of his mission, as Jeremy Black concluded: "Style meant competence, competence ensured style." Bond's outfits are worn with the confidence that is an essential part of his character. Costume designer Jany Temime, who dressed Bond in *Skyfall* notes: "Women find Bond's style so attractive because he always wears clothes that make him feel confident. . . . And it's that confidence, not the clothes, that's so attractive."[12] The same could be said about the way Bond uses his equipment. We never see a look of anxiety as he presses the button that brings the next gadget into action. We never see him struggling to make a machine work. He never has to make a call to a helpline or check the manual. Every machine works perfectly in a Bond film, and that alone makes them fantasies.

Fleming's novels were condemned for promoting "the cult of luxury for its own sake," and Jeremy Black describes Bond's world as "a wealth fantasy." He points out that Bond lives a life of luxury without the accompanying worries about money. Black sees this "cost-free" lifestyle as appealing both to affluent viewers and to those indulging in escapist fantasies, but he adds the telling point that these fantasies are "graspable."[13] The gadgets, travel, and fine dining the audience sees Bond enjoy onscreen, once the preserve of the very rich, have become more accessible. In 1967 one might not be able to afford a car with a Sony wireless communication inside like the one Bond drove in *You Only Live Twice,* but buying a Sony television or a Nikon camera gave consumers an opportunity to indulge in the advanced consumer technologies of the time.

In his study of Bond's watches Dell Deaton shows how they merged modernity and style. Fleming was a traditionalist, and nothing says traditional quality in timepieces better than Rolex—precise Swiss automatic movements encased in gold. But Bond on film has to move with the times, and his watches led the way into the "quartz revolution" in timepieces. The Seiko 0674 digital watch that Simon Winder marveled at in *The Spy Who Loved Me* marked a new era in wristwatches, and soon the 10-year-old Winder could afford a cheaper version. By the time of *Moonraker,* Bond has an LCD M354 Memory Bank Calendar watch, which again was available to consumers. Deaton has argued that the quartz revolution accomplished much more than introducing cheaper watches; it gave consumers more control over timekeeping than they had enjoyed before. The Seiko M354 not only featured the new LCD display but also provided the user with a calendar. In interpreting the quartz revolution as a consumer-driven technological development, Deaton argues that the consumers of watches wanted more functions and thus more control over timekeeping. This is important for Bond because he is often in a personal race against the clock, whether he is defusing an atomic bomb or making some time to dally with a woman, so watches are essential to his work, as well as signifying his modernity and style. Of the roughly 40 watches featured in the Bond films, more than half have been divers' watches. The chronometers he wears exclusively these days are derived from the complex timers worn by professional divers. These are functional pieces of equipment that divers must use to survive, and they look it: outsized, with a divers bezel and multiple enclosed crowns. Bond was the first to wear a divers' chronograph with a tuxedo, something no one other than Bond could have gotten away with.[14]

Much of Bond's derring-do involves strenuous activities like skiing, parachuting, diving, and climbing, which in the 1960s were becoming leisure-time activities. Diving had been confined primarily to military and commercial enterprises up to the 1950s, when scuba diving equipment came onto the market. During the war Jacques Cousteau had helped design the first generation of dive regulators, which controlled compressed air coming from tanks carried on the diver's back. The Aqua Lung was a simple piece of equipment that was easy to use, and this made it possible for an average person to dive. Although Bond always uses an Aqua Lung, it is the wrong equipment for undercover work because of the string of bubbles it leaves behind, but it looked more high-tech and manly than the rubberized rebreathing apparatus. Associating James Bond with scuba diving helped popularize the sport and gave it the aura of technological innovation and excitement that were core values of the Bond brand. As the sport grew in popularity in the late 1960s, more manufacturers of scuba gear, wet suits, and diving instruments entered the market, and the sport soon had its own magazine,

Scuba Diver, which began publication in 1967. One of the first celebrities featured on its cover was Sean Connery.

What connected James Bond so perfectly to the 1960s were the combined values of materialism and technological enthusiasm. Yet this was consumerism with a purpose. The exotic locales were a big part of what distinguished the Bond films from other spy films of the decade, for as Michael Denning points out, Bond is "the ideal tourist." None of the early Bond films would be complete without several scenes of him arriving in airports, checking in at luxury hotels, and ordering room service. Yet unlike those who were emulating his travels as commercial jet service brought the world closer to the affluent, Bond is on a mission, and this justifies his conspicuous consumption. Denning argues that "his tourism has an ostensible purpose, although the line between tourism and spying is a fine one."[15] Bond's consumer choices and his job were intrinsically linked; he wasn't wasting his time in casinos and fancy hotels because he was on active service. As Jeremy Black points out, Bond showed that "affluence and consumerism could be stylish and noble, rather than vulgarly materialistic."[16]

Reinventing Bond for twenty-first century audiences involved a return to a tougher and more callous character, the first signs of soul searching in his profession as a killer, and some important changes in his wardrobe. The "slim-fit" shirts and suits made by fashion designer Tom Ford for Daniel Craig in *Skyfall* reflected contemporary fashions and brought Bond up-to-date as a fashion icon. Daniel Craig as Bond did not make the market for Barbour jackets, which were already fashionable before *Skyfall,* but the model he wore in the film quickly sold out (a steal at $600) and could be found on eBay at premium prices. The "Fleming Effect" is now used by advertisers and marketers to describe the increase in sales of any item that James Bond wears or uses in the films.

Product placement has become an integral part of the latest Bond films. Range Rover, Sony, Tom Ford, and BMW lend their products, make sure their logos are clearly displayed, and produce their own tie-in advertising in print, media, and television commercials that have the same high production standards as the films. The 007 website currently lists 12 "brand partners" who have paid a pretty penny to have Bond use their products. It is said that Heineken paid $45 million to replace his vodka martinis with its beer in *Skyfall*—a sellout that outraged traditionalists. But with budgets for each film reaching up to $300 million, the money coming in from brand partners is essential if a film this expensive is to be made. The details of products that Fleming added to the plots, which he told Connery were "nothing but padding," have become essential to the Bond formula and the financing of the films.[17] Modernity was once the sole preserve of the gadgets, but now it is the merchandising that helps keep Bond up to date, as each device and vehicle is branded with the manufacturer's name. Bond's brand partners run unique advertising campaigns coordinated with the theatrical runs of the films, and this reinforces the cachet of newness. As Paul Stock points out, Bond is a "technologically progressive icon," whose dependence on his innovative gadgets gives him a technological advantage over his enemies. As Bond asserts his superiority with his spyware, these devices share in his triumphs—"the endorsed products are aligned with his successes, and with the debonair image of the super-spy." The gadgets that once defined Bond as a modernizing hero and won the day for the Secret Intelligence Service now have fetish powers that support merchandising. Stock concludes that Bond is "the hero of corporate capitalism, and the endorser of fetishised commodities."[18]

The Cult of the Individual

Lewis Gilbert, a director of several Bond films, pointed out that audiences "don't like change at all. . . . They like the pattern, the formula. I think that part of the charm of the Bond picture [is that] you know what you're going to get."[19] The howls of outrage that follow any suggestion that a future Bond might not be white, male, and English testifies to the importance of continuity in the Bond franchise. Audiences know they are going to see equipment that anticipates the future, and the films' producers recognize that they need to present Bond as a modern, tech-savvy hero, but he is also obliged to remain the same man who emerged at the end of World War II. In defending Western hegemony, Bond demonstrates the virtues of the early twentieth-century gentleman spy, yet as William Rees-Mogg of the *Times* has pointed out, there is a fundamental anachronism in "a high technology killer. . . . The Achilles of our thoroughly modern age should be a man born in the reign of King Edward VII."[20] Umberto Eco sees the literary Bond as a combination of nineteenth-century tradition and twentieth-century science fiction. The films might be in digitally high definition, but the villains, the torture, and the deaths remain "pure Gothic."[21] The dialectic between nostalgia and modernity in the Bond film franchise reflects Fleming's own ambiguity about technological change. Even as Bond becomes more machine-like, he owes his privileged place in popular culture to his resistance to machines, as symbolized in their ritual destruction at the climax of every film. As far as Fleming was concerned, Bond stood as a surrogate for resistance to change, and it is significant that the modernization of espionage, and the threat of replacement by machines that Fleming recognized in the 1950s, does not catch up to Bond until 2015 and *Spectre*. In this way Bond finds himself in the same no-man's-land as his Edwardian creator—stuck between the aspirations of two vastly different centuries, pressed to move forward while keeping to the values of the past. The plots and his equipment are driven by both cultural angst and technological change, but the world's most famous secret agent is there to protect the status quo in its many forms.

The one constant in the films and books is that Bond always triumphs against the machine. No matter how futuristic and dangerous the threat, Fleming's (and Bond's) faith in the resilience of human agency, in individual ingenuity and improvisation, still wins the day. In *The Spy Who Loved Me* it only takes two screwdrivers to disassemble the nuclear warhead of a Polaris missile and a few seconds spent examining a software manual to reprogram two Polaris missile launches—the first recorded instance of one-finger typing saving the world. The fight against evil has moved into the internet and cyberspace, against malicious hackers and digitally enhanced villains, but in the end tranquility is always restored by a hero "who takes power from the machine and hands it back to the human."[22] The Bond character reassures us that our own courage and ingenuity can overcome the anxieties of technological change, providing we have the right equipment.

NOTES

Introduction

1. James Chapman, *Licence to Thrill: A Cultural History of the James Bond Films* (New York: Columbia University Press, 2000), 16.

Chapter One. The Technological Enthusiasts

1. Alfred North Whitehead, *Science and the Modern World* (New York: Macmillan, 1925), 98.

2. Thomas P. Hughes, *American Genesis: A Century of Invention and Technological Enthusiasm* (New York: Penguin, 1989), 2–3, 7, 14, 297; Howard P. Segal, *Technological Utopianism in American Culture* (Chicago: University of Chicago Press, 1985); Lewis Mumford, *Technics and Civilization* (Chicago: University of Chicago Press, 2010).

3. Reese V. Jenkins, "George Eastman and the Coming of Industrial Research in America," in *Technology in America*, ed. Carroll W. Pursell Jr. (Cambridge, MA: MIT Press, 1981), 129–41, 141.

4. Winston S. Churchill, author's preface to *A Roving Commission: The Story of My Early Life* (New York: Scribner's, 1951), n.p.

5. Umberto Eco, "Narrative Structures in Fleming," in *The James Bond Phenomenon: A Critical Reader*, ed. Christoph Lindner (Manchester: Manchester University Press, 2009), 34–55, 53.

6. Churchill, *A Roving Commission*, 106.

7. H. G. Wells, *The War in the Air* (New York: Macmillan, 1908), 68, 8.

8. F. W. Winterbotham, *The Ultra Spy* (London: Macmillan, 1989), 4; de Havilland quoted in Sam Howe Verhovek, *Jet Age: The Comet, the 707, and the Race to Shrink the World* (New York: Penguin, 2010), 54.

9. Andrew Lycett, *Ian Fleming: The Man behind James Bond* (Atlanta: Turner, 1995), 52–53; Fergus Fleming, ed., *The Man with the Golden Typewriter: Ian Fleming's James Bond Letters* (New York: Bloomsbury, 2015), 173.

10. Quoted in John Pearson, *The Life of Ian Fleming* (New York: McGraw Hill, 1966), 71–72.

11. William Stevenson, *A Man Called Intrepid* (New York: Ballantine, 1976), 297.

12. Pearson, *The Life of Ian Fleming*, 29.

13. Eric Schatzberg, "Technik Comes to America: Changing Meanings of Technology before 1930," *Technology and Culture* 47, no 3 (2006): 486–512.

14. Eric Schatzberg, "The Struggle for Technology: Instrumentalism versus Culture," *Rethinking Technology* (blog), 8 August 2012, http://rethinktechnology.wordpress.com/2012/08/08/the-struggle-for-technology-instrumentalism-versus-culture; Mumford, *Technics*

and Civilization, 427. See also Robert L. Heilbroner, "Do Machines Make History?" *Technology and Culture* 8, no. 3 (1967): 335–45. For a detailed examination of the thought of Lewis Mumford see Donald L. Miller, *Lewis Mumford: A Life* (New York: Weidenfeld and Nicolson, 1989).

15. Taylor Downing, *Secret Warriors: The Spies, Scientists, and Code Breakers of World War I* (New York: Pegasus, 2013), 56–57; Winterbotham, *The Ultra Spy,* 87.

16. Lawrence Goldstone, *Birdmen: The Wright Brothers, Glenn Curtiss, and the Battle to Control the Skies* (New York: Ballantine, 2014), 133; Verhovek, *Jet Age,* 49–50.

17. Verhovek, *Jet Age,* 55–56.

18. Graham Farmelo, *Churchill's Bomb: How the United States Overtook Britain in the First Nuclear Arms Race* (New York: Basic Books, 2013), 24; Winston S. Churchill, *Thoughts and Adventures* (London: Odhams, 1932), 133–35.

19. Wells, *The War in the Air,* 103; Perry Miller cited in Hughes, *American Genesis,* 1; John Ellis, *The Social History of the Machine Gun* (Baltimore: Johns Hopkins University Press, 1986), 32–33.

20. Alan Judd, *The Quest for C: Mansfield Cumming and the Founding of the Secret Service* (New York: Harper Collins, 2000), 39–41, 47–48; Downing, *Secret Warriors,* 14.

21. Churchill, *A Roving Commission,* 182–94; Arnold quoted in Ellis, *Social History of the Machine Gun,* 86.

22. Wells, *The War in the Air,* 170; Winterbotham, *The Ultra Spy,* 9.

23. Lycett, *Ian Fleming,* 28. This quote is typical Fleming bravado because he was a poor horseman and his experience at Sandhurst's Cavalry School was "horrible." See Pearson, *The Life of Ian Fleming,* 21; Churchill, *A Roving Commission,* 66.

Chapter Two. The Secret Intelligence Service

1. Judd, *The Quest for C,* 104.

2. Quoted in Keith Jeffery, *The Secret History of MI6, 1909–1949* (New York: Penguin, 2010), 34.

3. Robert Baden-Powell, *My Adventures as a Spy* (1915; Stroud, Gloucestershire: Amberley, 2014), 77.

4. Arthur Conan Doyle, *The Adventure of the Bruce-Partington Plans,* in *The Annotated Sherlock Holmes,* ed. William Baring-Gould, 2 vols. (New York: Clarkson N. Potter, 1967), 2:432–47, 435–36. An annotation points out that the Royal Navy was not building submarines in 1895 (the first was a Holland design purchased from the United States in 1903) but that there was a new French submarine introduced in 1896, the *Morse,* which could have inspired the story (436).

5. Baden-Powell, *My Adventures as a Spy,* 34.

6. Tom Reiss, "Imagining the Worst," *New Yorker,* 28 November 2005, 106.

7. Nicholas Rankin, *A Genius for Deception: How Cunning Helped the British Win Two World Wars* (New York: Oxford, 2009), xiii; Churchill, *Thoughts and Adventures,* 87, 60.

8. Nigel West, ed., *MI5 in the Great War* (London: Biteback, 2014), 137, 29, 69, 80, 98, 125.

9. Jeffery, *The Secret History of MI6,* 122, 126.

10. Michael Peterson, *The Secret War: The Inside Story of the Code Makers and Code Breakers of World War II* (London: David and Charles, 2007), 23.

11. Downing, *Secret Warriors,* 107–8.

12. Nicholas C. Watkis, *The Western Front from the Air* (Stroud, Gloucestershire: History Press, 2013), 35.

13. Winterbotham, *The Ultra Spy,* 70; Peter Hart, *Aces Falling: Life above the Trenches, 1918* (London: Phoenix, 2007), 288; Watkis, *Western Front from the Air,* 35.

14. Basil Thomson, *Odd People: Hunting Spies in the First World War* (1922; London: Biteback, 2015), 37. Thomson added that "the longest-lived of the delusions was that of the night signalling" to Zeppelins (43).

15. Judd, *The Quest for C,* 412. Cumming had an interest in chemistry and an inquisitive mind. One way to recover a text written in secret ink was to coat it in iodine vapor. One of his agents, George Hill, wrote, "I shall never forget C's delight when the Chief Censor, Worthington, came in one day with the announcement that one of his staff had found out that semen would not respond to iodine vapour" (323).

16. Giles Milton, *Russian Roulette: How British Spies Thwarted Lenin's Plot for Global Revolution* (New York: Bloomsbury, 2013), 206; Andrew Cook, *Ace of Spies: The True Story of Sidney Reilly* (Stroud, UK: Tempus, 2004), 152, quoting SIS operative Paul Dukes, *Red Dusk and the Morrow: Adventures and Investigations in Red Russia* (London: Williams and Norgate, 1923), 9.

17. West, *MI5 in the Great War,* 248, 302, 406.

18. Milton, *Russian Roulette,* 71, 207, 86; Jeffery, *The Secret History of MI6,* 66–67, 168.

19. West, *MI5 in the Great War,* 35–37, 139, 223.

20. Calder Walton, *Empire of Secrets: British Intelligence, the Cold War, and the Twilight of Empire* (New York: Overlook, 2013), 8.

21. Jeffery, *The Secret History of MI6,* 141; Milton, *Russian Roulette,* 200.

22. Milton, *Russian Roulette,* 199.

Chapter Three. The Great War and the Threat of Modernity

1. John Buchan, *Buchan's War* (Stroud, Gloucestershire: Amberley, 2014), 155.

2. T. E. Lawrence, *Revolt in the Desert* (London: Taurus Parke, 2011), 240.

3. Churchill, *A Roving Commission,* 301.

4. Deneys Reitz, *Commando: A Boer Journal of the Boer War* (London: Faber and Faber, 1929), 23–25, 167, 314.

5. T. E. Lawrence, *Seven Pillars of Wisdom* (New York: Dell, 1962), 195–97.

6. Rankin, *A Genius for Deception,* 122.

7. Doyle, *The Adventure of the Bruce-Partington Plans,* 436.

8. West, *MI5 in the Great War,* 29; Churchill, *Thoughts and Adventures,* 87–88.

9. Jamie Prenatt and Mark Stille, *Axis Midget Submarines, 1939–45* (London: Osprey, 2014), 5.

10. Cecil Lewis, *Sagittarius Rising* (1936; London: Folio Society, 1998), 95; Cook, *Ace of Spies,* 142.

11. Arthur Gould Lee, *No Parachute* (London: Grub Street, 2013), 2, 221–22; James Hamilton-Patterson, *Marked for Death: The First War in the Air* (London; Head of Zeus, 2015), 26.

12. Peter Hart, *Bloody April: Slaughter in the Skies over Arras, 1917* (London: Cassell, 2005), 305.

13. Winterbotham, *The Ultra Spy,* 64.

14. Churchill, *Thoughts and Adventures,* 90.

15. Norman Franks, *Dogfight: Aerial Tactics and Aces of World War 1* (London: Greenhill, 2003), 29; Jack Herris and Bob Pearson, *Aircraft of World War I* (London: Amber Books, 2010), 120.

16. Downing, *Secret Warriors,* 354.

17. Evan Hadingham, "Germany's Titanic Triplane," *Aviation History* 6, no. 5 (2016): 12–13, 13.

18. Taylor Downing, *Spies in the Sky: The Secret Battle for Aerial Intelligence during World War II* (London: Little, Brown, 2011), 19–22.

19. Lee Kennett, *The First Air War, 1914–1918* (New York: Free Press, 1991), 10–11, 110–11.

20. Trevor Gibbons, "The Day World War One Came to Hull," 5 June 2015, BBC News, www.bbc.com/news/uk-england-humber-32917351; Ben Robinson, "World War One: How the German Zeppelins Wrought Terror," 4 August 2014, BBC News, www.bbc.com/news/uk-england-27517166.

21. Quoted in Patrick McGilligan, *Alfred Hitchcock: A Life in Darkness and Light* (New York: Harper Collins, 2003), 26.

22. R. V. Jones, *Most Secret War* (London: Wordsworth, 1998), 4.

23. Violet Hunt and Ford Madox Hueffer, *Zeppelin Nights: A London Entertainment* (London: John Lane, 1916), 1–2.

24. Wilbur Cross, *Zeppelins of World War I* (New York: Barnes and Noble, 1991), 1, 18, 20.

25. Michael MacDonagh, "Gotha Aeroplanes Bomb London, 7 July 1917," in *The Mammoth Book of Eyewitness World War I,* ed. Jon E. Lewis (New York: Carroll and Graf, 2004), 302–3; Churchill quote from House of Commons debate 30 July 1934, cited in Farmelo, *Churchill's Bomb,* 81.

26. Hunt and Hueffer, *Zeppelin Nights,* 16, 307.

Chapter Four. Imagining the Future

1. Goldstone, *Birdmen,* 22.

2. Michael Neufeld, "Weimar Culture and Futuristic Technology: The Rocketry and Spaceflight Fad in Germany, 1923–1933," *Technology and Culture* 31, no. 3 (1990): 725–52.

3. Rolf Aurich, Wolfgang Jacobsen, and Cornelius Schnauber, eds., *Fritz Lang: His Life and Work* (Berlin: Film Museum, 2001), 130–34.

4. Patrick McGilligan, *Fritz Lang: The Nature of the Beast* (New York: St Martin's, 1997), 329–30; Wayne Biddle, *Dark Side of the Moon: Wernher von Braun, the Third Reich, and the Space Race* (New York: Norton, 2009), 42–44, 48.

5. McGilligan, *Fritz Lang,* 144.

6. Wells, *The War in the Air,* 215–16.

7. Tom Gunning, *The Films of Fritz Lang: Allegories of Vision and Modernity* (London: BFI, 2000), 78–81.

8. H. G. Wells, *The War of the Worlds* (1898; Mineola, NY: Dover, 1997), 3.

9. George Orwell, "Boys Weeklies," in *George Orwell: Essays,* ed. John Carey (New York: Everyman's Library / Knopf, 2002), 201.

10. Harry Grindell Matthews, "The Death Power of Diabolical Rays," *New York Times,* 21 May 1924, 1:2; 3:4; "Tesla at 78 Bares New Death Beam," *New York Times,* 11 July 1934, 18:1–2.

11. Farmelo, *Churchill's Bomb,* 25; Winston S. Churchill, "Shall We All Commit Suicide?" in *Thoughts and Adventures,* 184–91, 189. He asks if explosives could be "guided automatically in flying machines by wireless or other rays, without a human pilot, in a ceaseless procession upon a hostile city" (189).

12. David Edgerton, *Britain's War Machine: Weapons, Resources, and Experts in the Second World War* (New York: Oxford University Press, 2011), 14, 35.

13. Quoted in Stephen Budiansky, *Blackett's War* (New York: Knopf, 2013), 71, 70.

14. Winterbotham, *The Ultra Spy,* 146–47, 160. Section IV of SIS was the air section, which joined Sections I (political), II (which liaised with the army), and III (which served the navy).

15. Tom Bower, *The Paperclip Conspiracy: The Hunt for Nazi Scientists* (Boston: Little, Brown, 1987), 22. This committee was severely hampered by the rivalry and disagreements between Tizard and Lindemann; see Farmelo, *Churchill's Bomb,* 84–86.

16. Bower, The Paperclip Conspiracy, 29.

17. Ibid., 27.

Chapter Five. Spy Films

1. Lycett, *Ian Fleming,* 32; Ian Fleming, *Thrilling Cities* (New York: Signet, 1965), 32.

2. McGilligan, *Fritz Lang,* 136.

3. Pearson, *The Life of Ian Fleming,* 81.

4. Milton, *Russian Roulette,* 99.

5. Pepita Reilly, *Britain's Master Spy: The Adventures of Sidney Reilly* (New York: Harper and Bros., 1932), xii; Milton, *Russian Roulette,* frontispiece, 81, 108.

6. Pearson, *The Life of Ian Fleming,* 11.

7. A MacGuffin is a plot device in the form of some goal or desired object that the protagonist pursues but is not important to the plot and usually not explained. Hitchcock defined it as "the unknown plot objective which you need not choose until the story planning was complete."

8. John Buchan, *The Thirty-Nine Steps* (New York: Doran, 1915), 197.

9. William Stevenson, *Spymistress* (New York: Arcade, 2007), front cover.

10. Aurich, Jacobsen, and Schnauber, *Fritz Lang,* 384.

11. Simon Winder, *The Man Who Saved Britain* (London: Picador, 2006), 131–32.

Chapter Six. Ian Fleming, Intelligence Officer

1. Pearson, *The Life of Ian Fleming,* 88.

2. Lycett, *Ian Fleming,* 94–95; Fleming, *Thrilling Cities,* 122.

3. Pearson, *The Life of Ian Fleming,* 93.

4. Stephen Bull, ed., *The Secret Agent's Pocket Manual, 1939–1945* (based on documents published by SOE and the OSS) (London: Conway, 2009), 27.

5. Winterbotham, *The Ultra Spy,* 189–91.

6. Lycett, *Ian Fleming,* 106–7.

7. Downing, *Spies in the Sky,* 38.

8. Winterbotham, *The Ultra Spy,* 190; Jones, *Most Secret War,* 130.

9. Downing, *Spies in the Sky,* 31.

10. Ibid., 73, 101.

11. Fleming, *Thrilling Cities,* 54.

12. Ben Macintyre, *Double Cross: The True Story of the D-Day Spies* (New York: Crown, 2012), 148.

13. Downing, *Spies in the Sky,* 251.

14. Macintyre, *Double Cross,* 80.

15. Stuart Macrae, *Winston Churchill's Toyshop* (Stroud, Gloucestershire: Amberley, 1972), 134.

16. Nigel West, *MI6: British Intelligence Service Operations, 1909–1945* (New York: Random House, 1983), 62.

17. Macrae, *Churchill's Toyshop,* 21, 88.

18. This refers to the poison pills issued to all agents; see E. H. Cookridge, *Inside SOE: The Story of Special Operations in Western Europe, 1940–1945* (London: Arthur Barker, 1966), 88.

19. Fredric Boyce and Douglas Everett, *SOE: The Scientific Secrets* (Stroud, Gloucestershire: History Press, 2013), 46, 97–98.

20. Cookridge, *Inside SOE,* 90–92; Roderick Bailey, *Forgotten Voices of the Secret War: An Inside History of Special Operations during the Second World War* (London: Ebury, 2009), 69.

21. M. R. D. Foot, *SOE: The Special Operations Executive, 1940–1946* (London: Pimlico, 1999), 91.

22. Edgerton, *Britain's War Machine,* 234–35.

23. Macrae, *Churchill's Toyshop,* 84.

Chapter Seven. Equipment

1. Charles Fraser-Smith, with Gerald McKnight and Sandy Lesberg, *The Secret War of Charles Fraser-Smith* (London: Michael Joseph, 1981), 128–29.

2. Macrae, *Winston Churchill's Toyshop,* 172; Pearson, *The Life of Ian Fleming,* 102–3. Fleming claimed the pen could also be charged with poison gas.

3. Fraser-Smith, *Secret War of Charles Fraser-Smith,* 21–24, 37–40; Winterbotham, *The Ultra Spy,* 147. Winterbotham had a secret source within Goering's staff before the war, receiving from him envelopes containing Photostats of Luftwaffe documents.

4. Bettina Stangneth, *Eichmann before Jerusalem: The Unexamined Life of a Mass Murderer* (New York: Vintage, 2015), 6–7.

5. Fraser-Smith, *Secret War of Charles Fraser-Smith,* 128.

6. Jeffery, *The Secret History of MI6,* 485.

7. Macrae, *Winston Churchill's Toyshop,* 17–20.

8. Bailey, *Forgotten Voices,* 68, 150; Macrae, *Winston Churchill's Toyshop,* 100.

9. Macrae, *Winston Churchill's Toyshop,* 105–7.

10. Ibid., 119; Nigel Cawthorne, *Reaping the Whirlwind: The German and Japanese Experience of World War II* (Newton Abbot, UK: David and Charles, 2007), 69–71.

11. Macrae, *Winston Churchill's Toyshop*, 134.

12. Abwehr technicians thought them to be junk. See James Hayward, *Hitler's Spy: The True Story of Arthur Owens, Double Agent Snow* (New York: Simon and Schuster, 2014), 30–31.

13. Bailey, *Forgotten Voices*, 65.

14. Foot, *SOE*, 12, 146; Prenatt and Stille, *Axis Midget Submarines*, 23.

15. Fleming, *Thrilling Cities*, 205–7.

16. Ibid., 103; Prenatt and Stille, *Axis Midget Submarines*, 23; quoted in Nicholas Rankin, Ian *Fleming's Commandos: The Story of 30 Assault Unit in World War II (New York: Pegasus, 2014)*, 146.

Chapter Eight. Irregular Warriors

1. Cookridge, *Inside SOE*, 529; Phillip Knightley, *The Second Oldest Profession: Spies and Spying in the Twentieth Century* (New York: Norton, 1987), 122; Kim Philby, *My Silent War: The Autobiography of a Spy* (New York: Modern Library, 2002), 12, 82.

2. Foot, *SOE*, 241. David Hare's play and film *Plenty* reflects on this and the difficulties of coming to terms with the boring reality of postwar life.

3. Bailey, *Forgotten Voices*, 99.

4. Stephen Dorril, *MI6: Inside the Covert World of Her Majesty's Secret Intelligence Service* (New York: Free Press, 2000), 250; West, *MI6*, 100–102; Winterbotham, *The Ultra Spy*, 196.

5. Ben Macintyre, *For Your Eyes Only: Ian Fleming+James Bond* (London: Bloomsbury and Imperial War Museum, 2008), 68, 107.

6. Foot, *SOE*, 12, 302.

7. Stephen E. Ambrose, *Band of Brothers* (New York: Pocket Books, 1992), 20.

8. Flint Whitlock, *If Chaos Reigns: Allied Airborne Forces on D-Day* (Philadelphia: Casemate, 2011), 49, 29; Evelyn Waugh, *Put Out More Flags* (Boston: Little Brown, 1977), 279–80.

9. Tim Lynch, *Operation Market Garden* (Stroud, Gloucestershire: Spellmount, 2011), 84–85; Bailey, *Forgotten Voices*, 32.

10. Pearson, *The Life of Ian Fleming*, 13; Stevenson, *A Man Called Intrepid*, 209.

11. Bailey, *Forgotten Voices*, 43, 60; Cookridge, *Inside SOE*, 78. Not everyone on the course joined in this charade: "Oh, Peter [Folliss], for fuck's sake, let's get back to bed."

12. Bailey, *Forgotten Voices*, 58; Donald Thomas, *The Enemy Within: Hucksters, Racketeers, Deserters, and Civilians during the Second World War* (New York: New York University Press, 2003), 327. Gentleman Johnny went on to carry out secret, safecracking missions in North Africa, Italy, and Germany.

13. Bailey, *Forgotten Voices*, 47; Fleming, *Thrilling Cities*, 49; W. E. Fairbairn, *All-In Fighting* (Uckfield, Sussex: Naval and Military Press, facsimile of 1942 edition), 7–9.

14. Bailey, *Forgotten Voices*, 47–48; Bull, *The Secret Agent's Pocket Manual*, 63–64; Cookridge, *Inside SOE*, 76.

15. Foot, *SOE*, 118.

16. Bailey, *Forgotten Voices*, 106, 110.

17. Eleanor Pelrine and Dennis Pelrine, *Ian Fleming: The Man with the Golden Pen* (Wilmington, DE: Swan, 1966), 116–17.

18. Pearson, *The Life of Ian Fleming*, 37–38.

Chapter Nine. The Treasure Hunt

1. Rankin, *Fleming's Commandos*, 131.

2. Pearson, *The Life of Ian Fleming*, 213, 203.

3. Bailey, *Forgotten Voices*, 77.

4. Ibid., 80–81.

5. Rankin, *Fleming's Commandos*, 159–61.

6. Dorril, *MI6*, 137; Rankin, *Fleming's Commandos*, 181; Tony Rennell, "The Skirt Chasing Killers Inspired 007," *Daily Mail*, 12 October 2011, www.dailymail.co.uk/news/article-2048506.html.

7. Rankin, *Fleming's Commandos*, 220; Fleming, *Thrilling Cities*, 122.

8. Neal Bascomb, *Hunting Eichmann* (Boston: Houghton Mifflin, 2009), 16; Thomas Harding, *Hanns and Rudolf* (New York: Simon and Schuster, 2013), 151.

9. Chapman Pincher, *Too Secret, Too Long* (New York: St Martin's, 1984), 90.

10. Barry Day, ed., *The Noel Coward Reader* (New York: Knopf, 2010), 366.

11. Bailey, *Forgotten Voices*, 136, 138.

12. Roger Moorhouse, *Berlin at War* (New York: Basic Books, 2010), 344.

13. Rankin, *Fleming's Commandos*, 315.

14. Bower, *The Paperclip Conspiracy*, 121; Norman Longmate, *Hitler's Rockets: The Story of the V-2s* (London: Frontline, 2009), 201, 203.

15. Longmate, *Hitler's Rockets*, 132–33; Jones, *Most Secret War*, 457, 459.

16. Longmate, *Hitler's Rockets*, 201, 203; Jones, *Most Secret War*, 455–57.

17. Rankin, *Fleming's Commandos*, 236.

18. Roger Howard, *Operation Damocles: Israel's Secret War against Hitler's Scientists, 1951–1967* (New York: Pegasus, 2013), 149, 191, 198.

19. Fleming, *Thrilling Cities*, 126.

20. Bower, *The Paperclip Conspiracy*, 67, 113.

21. *Secrets of Hitler's Wonder Weapons* (Nugus/Martin, 2002), via the History Channel.

22. See, e.g., ibid.; *Iron Sky* (Eone Films, 2012); *Alien Contact: Nazi UFOs* (Reality Films, 2016). Nazi zombies can be seen in the *Dead Snow* (Euforia Film, 2009) and *Zombie Massacre* (Meathead Records, 2002) films. Italian producers made many "Nazisploitation" films in the 1970s, including the imaginative *Caligula Reincarnated as Hitler* (Cine Lu.Ce., 1977).

Chapter Ten. Nuclear Anxieties

1. Yanek Mieczkowski, *Eisenhower's Sputnik Moment: The Race for Space and World Prestige* (Ithaca, NY: Cornell University Press, 2013), 266.

2. Winston S. Churchill, "Fifty Years Hence," in *Thoughts and Adventures*, 203–14, 208.

3. Melvin Kranzberg, "Technology and History: Kranzberg's Laws," *Technology and Culture* 27, no. 3 (1986): 544–60.

4. Paul Boyer, *By the Bomb's Early Light: American Thought and Culture at the Dawn of the Nuclear Age* (New York: Pantheon, 1985), 299–300.

5. Michael Kerrigan, *Cold War Plans That Never Happened, 1945–91* (London: Amber, 2012), 50–51; Gregory Benford, *Popular Mechanics: The Amazing Weapons That Never Were: Robots, Flying Tanks, and Other Machines of War* (New York: Hearst, 2012), 37–42.

6. Farmelo, *Churchill's Bomb,* 3, 4, 408.

7. Norman Kagan, *The Cinema of Stanley Kubrick* (New York: Continuum, 1995), 111.

8. Peter Lamont, production designer, "Designing Bond," *Octopussy: Two-Disc Ultimate Edition,* disc 2, "Special Features" (Los Angeles, CA: MGM, 2008), DVD.

9. Robert Harris and Jeremy Paxman, *A Higher Form of Killing: The Secret Story of Chemical and Biological Warfare* (New York: Hill and Wang, 1990), 53–54.

Chapter Eleven. Gadgets

1. John Cork and Bruce Scivally, *James Bond: The Legacy* (New York: Harry N. Abrams, 2002), 315, 38.

2. Ken Adam, production designer, "Commentary," *Dr. No,* dir. Terence Young (1962; Los Angeles, CA: MGM, 2007), DVD.

3. Christoph Lindner, "Criminal Vision and the Ideology of Detection in the 007 Series," in *The James Bond Phenomenon: A Critical Reader,* ed. Christoph Lindner (Manchester: Manchester University Press, 2003), 76–88, 82; Ian Fleming, *Doctor No* (New York: Macmillan, 1958), 142–43.

4. Quoted in Cork and Scivally, *James Bond,* 68.

5. Cited in ibid., 78; David Thomson, *The Big Screen: The Story of the Movies* (New York: Farrar, Straus and Giroux, 2012), 366.

6. Ben Macintyre, *A Spy among Friends: Kim Philby and the Great Betrayal* (New York: Broadway, 2015), 187.

7. Quoted in Chapman, *Licence to Thrill,* 59.

8. "Robert Fulton's Skyhook and Operation Coldfeet," CIA website, https://www.cia.gov/library/center-for-the-study-of-intelligence/csi-publications/csi-studies/studies/95unclass/Leary.html.

9. Lisa Davis, "36,000 Feet under the Sea, " *SF Weekly,* 10 June 1998, www.sfweekly.com/sanfrancisco/36000-feet-under-the-sea/Content?oid=2135357.

10. J. Hoberman, *The Dream Life: Movies, Media, and the Mythology of the Sixties* (New York: New Press, 2003), 61–62; Thomas Powers, *The Man Who Kept the Secrets: Richard Helms and the CIA* (New York: Pocket Books, 1979), 190–91.

11. Gordon Corera, *The Art of Betrayal: The Secret History of MI6* (New York: Pegasus, 2014), 180, 223; The character of Max Otto von Stierlitz was the hero of a 12-part series, *Seventeen Moments of Spring,* first broadcast in the Soviet Union in 1973, in which he infiltrates the German High Command and undermines peace negotiations between the Nazis and the Americans. The TV series was commissioned by the KGB and filmed under its supervision.

12. Gordievsky quoted in Cork and Scivally, *James Bond,* 50; Wallace quoted in Danny Biederman, *The Incredible World of Spy-Fi* (San Francisco: Chronicle, 2004), 6–7.

13. Philby, *My Silent War,* 56; Genrikh Borovik and Phillip Knightley, eds., *The Philby Files: The Secret Life of Master Spy Kim Philby* (Boston: Little, Brown, 1994), 247–48; Gavin Esler, "How Nazi Adolf Eichmann's Holocaust Trial Unified Israel," BBC Radio 4, 6 April 2011, www.bbc.com/news/world-12912527; Leonard LeSchack, *Secret Weapons of Spies* (World Media Rights Production, 2012), via the History Channel.

14. Corera, *The Art of Betrayal,* 174.

15. Fleming quoted in Macintyre, *For Your Eyes Only,* 107.

Chapter Twelve. Guns

1. Anthony Lane, "Human Bondage," *New Yorker,* 16 November 2015, 97.

2. Quoted in Cork and Scivally, *James Bond,* 49.

3. Pearson, *The Life of Ian Fleming,* 204; Fergus Fleming, *The Man with the Golden Typewriter,* 32.

4. Lycett, *Ian Fleming,* 301, 298–99.

5. Gene Gangarosa Jr., *The Walther Handgun Story* (Wayne, NJ: Stoeger, 1999), 102–3. Adolf Hitler used a PPK to end his life on April 30, 1945.

6. David Chasman quoted in Laurent Bouzereau, *The Art of Bond* (New York: Abrams, 2006), 227. The creative staff were given license to use their imaginations. The Bond films were "a special effects man's dream. You've got your big explosions, you've got your massive hydraulic lifts, you've got your teensy-weensy gadgets." Chris Corbould quoted in Cork and Scivally, *James Bond,* 311.

7. Lycett, *Ian Fleming,* 300.

8. Foot, *SOE,* 9–10.

9. Bailey, *Forgotten Voices,* 213.

10. The "long barrel" is difficult to explain because this was not a feature of the weapons issued to the military and completely unsuitable for a secret agent who has to pull it out of his inside pocket. Perhaps Fleming got the idea from the Colt Python, which has an eight-inch barrel, nearly double the length of the Peacemaker. See A. E. Hartnik, *Complete Encyclopedia of Pistols and Revolvers* (Edison, NJ: Chartwell, 2003), 113.

11. Boyce and Everett, *SOE,* 102–9.

12. Dorril, *MI6,* 87, 613, 683.

13. Bailey, *Forgotten Voices,* 19.

14. Foot, *SOE,* 102.

15. Boyce and Everett, *SOE,* 53.

16. Martin Willis, "Hard-Wear: The Millennium, Technology, and Brosnan's Bond," in *The James Bond Phenomenon: A Critical Reader,* ed. Christoph Lindner (Manchester: Manchester University Press, 2003), 169–83, 170.

Chapter Thirteen. The Special Relationship and the Cold War

1. Stevenson, *A Man Called Intrepid,* 138–39, 178, 297; Fleming, *Thrilling Cities,* 77.

2. On Fleming's note to Cornelius Ryan, who had just begun work on Donovan's biography, see Pearson, *The Life of Ian Fleming,* 158–59. Similarly, Stevenson's *A Man Called Intrepid* puts Fleming and William Stephenson at the center of Anglo-American intelli-

gence cooperation, but it was clearly more than the work of just these two men. Stephen Dorril credits BSC's Charles Ellis and Walter Bell. See Patrick Beesley, *Very Special Intelligence: The Story of the Admiralty's Operational Intelligence Center, 1939–1945* (New York: Doubleday, 1978), 113; Dorril, *MI6*, 51.

3. Douglas Waller, *Wild Bill Donovan: The Spymaster Who Created the OSS and Modern American Espionage* (New York: Free Press, 2012), 77, 88. The writer of that note to the president was his son Capt. James Roosevelt of the US Marines, 13 January 1942; see Jon T. Hoffman, *Once a Legend: "Red Mike" Edson of the Marine Raiders* (Novato, CA: Presidio, 2000), 151.

4. Stanley P. Lovell, *Of Spies and Stratagems* (New York: Prentice Hall, 1963), 17.

5. R. Harris Smith, *OSS: The Secret History of America's First Central Intelligence Agency* (Berkeley: University of California Press, 1972), 33; Dorril, *MI6*, 51.

6. Waller, *Wild Bill Donovan*, 101–3. Lovell brought in experts from the Surgeon General's Office and Cornell University Medical School to continue these experiments.

7. H. Montgomery Hyde, *Room 3603* (New York: Lyons, 1962), 172; Smith, *OSS*, 44.

8. Stevenson, *Spymistress*, 132, 265; Robert Wallace and H. Keith Melton, *Spycraft* (London: Bantam, 2009), 8–9.

9. Waller, *Wild Bill Donovan*, 223, 233.

10. Quoted in Wallace and Melton, *Spycraft*, 7.

11. Dorril, *MI6*, 711.

12. Waller, *Wild Bill Donovan*, 188–89.

13. Smith, *OSS*, 34: Dorril, *MI6*, 51; Philby, *My Silent War*, 78.

14. Cookridge, *Inside SOE*, 63.

15. Philby, *My Silent War*, 73.

16. Jerrold Schecter and Leona Schecter, *Sacred Secrets: How Soviet Intelligence Operations Changed American History* (Washington: Brassey's, 2003), 117–18; Jeffrey T. Richelson, *The Wizards of Langley: Inside the CIA's Directorate of Science and Technology* (Boulder, CO: Westview, 2001), 3.

17. Dorril, *MI6*, 396. The split in opinion in London generally followed MI5 versus MI6 lines, the former believing him guilty, the latter innocent. Philby's friend James Angleton, of CIA counterintelligence, supported him and proclaimed his innocence to the very end, while the FBI claimed all along that they suspected Philby of being a spy.

18. Dorril, *MI6*, 709–10. George Blake was sprung from Wormwood Scrubs prison in 1966, which was another embarrassment. He reappeared in Moscow and spent some time with Kim Philby. Blake affirmed that Nicholas Elliott "was especially sent out to Beirut to warn him [Philby] not to return to England . . . [and that] further scandals had to be avoided at all costs." George Blake, *No Other Choice: An Autobiography* (New York: Simon and Schuster, 1990), 211.

19. Andrew Boyle, *The Fourth Man* (New York: Dial, 179), 444.

20. Philby, *My Silent War*, epigraph, 10.

21. Verhovek, *Jet Age*, 9–10, 20; Fleming, *Thrilling Cities*, 12, 63.

22. Macintyre, *A Spy among Friends*, 107; Blake, *No Other Choice*, 100; Pearson, *The Life of Ian Fleming*, 172.

23. Peter Wright, *Spycatcher: The Candid Autobiography of a Senior Intelligence Officer* (New York: Dell, 1987), 24, 46, 90–91; Chester Cooper, *The Lion's Last Roar* (New York: Harper and Row, 1978), 70.

24. Wright, *Spycatcher,* 28, 30. The cutbacks of 1945, "this year of madness" as R. V. Jones called it, created "the age of abdication that has paralysed us since 1945." Jones, *Most Secret War,* 256.

25. John Le Carré, *Call for the Dead* (New York: Fall River, 2007), 24–25.

26. Philby, *My Silent War,* 40.

27. Corera, *The Art of Betrayal,* 71, 78.

28. Philby, *My Silent War,* 110; Beesley, *Very Special Intelligence,* 37.

29. Rankin, *Fleming's Commandos,* 313.

Chapter Fourteen. The Technological Revolution

1. Philby recording, broadcast on BBC Radio 4, 4 April 2016, www.bbc.co.uk/programmes/b076v1zq.

2. Wright, *Spycatcher,* 44.

3. Ibid., 24; see also Nigel West, *Molehunt: Searching for Soviet Spies in British Intelligence* (New York: Berkley, 1991), xvii.

4. Walton, *Empire of Secrets,* 8, 33.

5. West, *MI6,* 213; Hyde, *Room 3603,* 175–76; Philby, *My Silent War,* 177.

6. Blake, *No Other Choice,* 94; Philby, *My Silent War,* 134.

7. Len Deighton, *The Ipcress File* (London: Harper, 2011), 89–90.

8. Dorril, *MI6,* 57.

9. Wallace and Melton, *Spycraft,* 48.

10. Jeffery, *The Secret History of MI6,* 644–45.

11. Oleg Kalugin, *Spymaster: My Thirty-Two Years in Intelligence and Espionage against the West* (New York: Basic Books, 2009), 303–4.

12. Richelson, *The Wizards of Langley,* 2–4.

13. Mieczkowski, *Eisenhower's Sputnik Moment,* 16; Richelson, *The Wizards of Langley,* 23–26. The KH-2's camera provided resolution down to 25 feet, versus the KH-1's 40 feet.

14. Thomas Graham and Keith Hansen, *Spy Satellites* (Seattle: University of Washington Press, 2007), 37–39; Mieczkowski, *Eisenhower's Sputnik Moment,* 215.

15. Wallace and Melton, *Spycraft,* 169, 176, 63.

16. Ibid., 164–65; Wright, *Spycatcher,* 25, 146.

17. Tennent H. Bagley, *Spy Wars: Moles, Mysteries, and Deadly Games* (New Haven, CT: Yale University Press, 2007), 15.

18. Schecter and Schecter, *Sacred Secrets,* 268–69.

19. Wallace and Melton, *Spycraft,* 89–92, 101–4.

20. Ibid., 59, 89.

21. Richelson, *The Wizards of Langley,* 5.

22. Lycett, *Ian Fleming,* 362.

23. Schecter and Schecter, *Sacred Secrets,* 272; Corera, *The Art of Betrayal,* 174. Wynne and Penkovsky used the old spy-film trick of playing loud music at one of their clandestine

meetings to confound Soviet listening devices and were shocked when the Russians played them audible recordings of this meeting.

24. Dorril, *MI6,* 128–30, 523–26. In *From Russia with Love* the SIS station chief in Istanbul has a tunnel that runs under KGB headquarters but only a periscope to view the proceedings. This device provides an image of the target but no sound—a significant disadvantage in an espionage environment more and more focused on aural intelligence. Bond employs this piece of World War II technology to check out Tatiana Romanova's legs but gets no other intelligence from it.

25. Blake, *No Other Choice,* 100; Wright, *Spycatcher,* 61.

26. Winder, *The Man Who Saved Britain,* 13.

27. Lycett, *Ian Fleming,* 362.

Chapter Fifteen. Into the Future

1. Simon Winder, "Why James Bond Is a Religion," *Guardian,* 3 October 2015, www.theguardian.com/books/2015/oct/03/why-james-bond-is-a-religion-spectre-007-ian-fleming.

2. Mieczkowski, *Eisenhower's Sputnik Moment,* 124.

3. James Hamilton-Patterson, *Empire of the Clouds: When Britain's Aircraft Ruled the World* (London: Faber and Faber, 2010), 143, 5.

4. Kevin Rawlinson, "Sir Ken Adam, Oscar-Winning Production Designer, Dies Aged 95," *Guardian,* 10 March 2016, www.theguardian.com/film/2016/mar/10/sir-ken-adam-oscar-winning-production-designer-dies-aged-95.

5. "Wing Commander Ken Wallis," obituary, *Daily Telegraph,* 24 June 2016, www.telegraph.co.uk/news/obituaries/10286869/Wing-Commander-Kenneth-Wallis.html.

6. Len Deighton, *London Match* (New York: Knopf, 1985), 178.

7. Chapman, *Licence to Thrill,* 103; Barry Parker, *Death Rays, Jet Packs, Stunts, and Supercars: The Fantastic Physics of Film's Most Celebrated Secret Agent* (Baltimore: Johns Hopkins University Press, 2005), 60. Laser beams exerted a powerful fascination for Broccoli and Saltzman, and their sharp, colored lights appear in many Bond films, but Parker points out that they are not visible to the naked eye, so what we see on film is just a special effect.

8. Quoted in Cork and Scivally, *James Bond,* 100.

9. Willis, "Hard-Wear," 180.

10. Tony Bennett and Janet Woollacott, *Bond and Beyond: The Political Career of a Popular Hero* (London: Macmillan, 1987), 132.

11. Neil deGrasse Tyson, "Bond Gadgets Stand Test of Time (But Not Physics)," interview by David Greene, NPR's *Morning Edition,* 4 October 2012, www.npr.org/2012/10/04/162182129/bond-gadgets-stand-test-of-time-but-not-physics.

12. Parker, *Death Rays,* 85.

13. See Dell Deaton, "The Time When James Bond Did Not Wear a Wristwatch," *James Bond Watches* (blog), 25 October 2015, http://jamesbondwatchesblog.com/2015/10/25/the-time-when-james-bond-did-not-wear-a-wristwatch-part-3/; and "James Bond Watches: 'The List' of EON Productions Movie Wristwatches," www.jamesbondwatches.com/the-list-eon-movies.htm.

14. Ian Fleming, *Chitty Chitty Bang Bang* (Somerville, MA: Candlewick, 2013), 28–29. Fleming writes that "all bits of machinery that people love are made into females" (57).

15. Fergus Fleming, *The Man with the Golden Typewriter,* 288.

16. Ibid., 121.

17. Eco, "Narrative Structures in Fleming," 35.

18. Patrick O'Donnell, "James Bond, Cyborg Aristocrat," in *Ian Fleming and James Bond: The Cultural Politics of 007,* ed. Edward P. Comentale, Stephen Watt, and Skip Willman (Bloomington: Indiana University Press, 2005), 55–68, 63.

19. Colleen M. Tremonte and Linda Racioppi, "Body Politics and *Casino Royale:* Gender and (Inter)national Security," in *The James Bond Phenomenon: A Critical Reader,* ed. Christoph Lindner (Manchester: Manchester University Press, 2003), 184–201, 195.

Chapter Sixteen. Keeping Up with the Times

1. Cynthia Baron, "*Doctor No:* Bonding Britishness to Racial Sovereignty," in *The James Bond Phenomenon: A Critical Reader,* ed. Christoph Lindner (Manchester: Manchester University Press, 2003), 153–68, 162.

2. Christopher Hitchens, "Bottoms Up," *Atlantic,* April 2006, www.theatlantic.com/magazine/archive/2006/04/bottoms-up/304719.

3. Cited in Chapman, *Licence to Thrill,* 118–19.

4. Robert L. Heilbroner, "Do Machines Make History?" *Technology and Culture* 8, no. 3 (1967): 335–45, 345; Yaron Ezrahi, Everett Mendelsohn, and Howard P. Segal, eds., *Technology, Pessimism, and Postmodernism* (Amherst, MA.: University of Massachusetts Press, 1994), 2–4; Leo Marx, "The Idea of 'Technology' and Postmodern Pessimism," in *Does Technology Drive History?* ed. Merritt Roe Smith and Leo Marx (Cambridge, MA: MIT Press, 1994).

5. See Matthew Wisnioski, *Engineers for Change: Competing Visions of Technology in 1960s America* (Cambridge, MA: MIT Press, 2012).

6. Willis, "Hard-Wear," 170–73.

7. Winder, *The Man Who Saved Britain,* 281–82.

8. Hitchens, "Bottoms Up"; Aaron Jaffe, "James Bond, Meta-Brand," in *Ian Fleming and James Bond: The Cultural Politics of 007,* ed. Edward P. Comentale, Stephen Watt, and Skip Willman (Bloomington: Indiana University Press, 2005), 87–106, 94.

9. Fergus Fleming, *The Man with the Golden Typewriter,* 258; Mark Harris, *Pictures at a Revolution: Five Movies and the Birth of a New Hollywood* (New York: Penguin, 2008), 330.

10. Downing, *Secret Warriors,* 203; Foot, *SOE,* 107.

11. Clair Hines, "Entertainment for Men: Uncovering the *Playboy* Bond," in *The James Bond Phenomenon: A Critical Reader,* ed. Christoph Lindner (Manchester: Manchester University Press, 2003), 89–108, 91, 97–100.

12. Jeremy Black, *The Politics of James Bond: From Fleming's Novels to the Big Screen* (Westport, CT: Praeger, 2001), 98; Temime quoted in Lee Kynaston, "Six Ways to Look as Stylish as James Bond," *Daily Telegraph,* 26 October 2015, www.telegraph.co.uk/men/fashion-and-style/11951575/Six-ways-to-look-as-stylish-as-James-Bond.html.

13. Fergus Fleming, *The Man with the Golden Typewriter,* 184; Black, *The Politics of James Bond,* 211.

14. Deaton, "The Time When James Bond Did Not Wear a Wristwatch." In 1995 Bond ditched his Rolex for a Seamaster chronometer, made by Eon's marketing partner Omega. A special edition Omega "Spectre 300 Seamaster" was conspicuous in *Spectre.*

15. Michael Denning, "Licensed to Look: James Bond and the Heroism of Consumption," in *The James Bond Phenomenon: A Critical Reader,* ed. Christoph Lindner (Manchester: Manchester University Press, 2003), 56–75, 66.

16. Black, *The Politics of James Bond,* 211.

17. Jeff Smith, "Creating a Bond Market," in *The James Bond Phenomenon: A Critical Reader,* ed. Christoph Lindner (Manchester: Manchester University Press, 2003), 136–52, 139; Connery quoted in "Bondomania," *Time,* 11 June 1965, 59; "Bond v Bond: The Return of 007," *Economist,* 26 October 2015, www.economist.com/blogs/graphicdetail/2015/10/daily-chart-13.

18. Paul Stock, "Dial 'M' for Metonym: Universal Exports, M's Office Space and Empire," in *The James Bond Phenomenon: A Critical Reader,* ed. Christoph Lindner (Manchester: Manchester University Press, 2003), 251–67, 255, 257.

19. Quoted in Janet Woollacott, "The James Bond Films: Conditions of Production," in *The James Bond Phenomenon: A Critical Reader,* ed. Christoph Lindner (Manchester: Manchester University Press, 2003), 117–35, 119.

20. William Rees-Mogg, *Times* (London), 29 December 1997, quoted in Black, *The Politics of James Bond,* 201.

21. Eco, "Narrative Structures in Fleming," 49.

22. Willis, "Hard-Wear," 176.

INDEX

Abwehr, 71, 81, 97, 99
Adam, Ken, 114, 121, 172; and "Adam look," 167, 169; and gadgets, 126, 171; and sixties modernism, 119, 170. *See also* Eon Productions
aerial reconnaissance, 22–23, 27, 36, 65–67, 156; in artillery spotting, 31–32; U-2 spy planes, 158, 160; via satellite, 158–59, 160, 162, 171
aerial reconnaissance cameras: A-type, 22; C-type, 22; Dubowsky cassette system, 35; E-type, 23; F-24, 36; Fairchild, 65; L-type, 23
American intelligence, 144–47
Armalite firearms, 120, 137
Ashenden; or, The British Agent (Maugham), 55
assassination, 135, 136, 137–38; and single-shot firearms, 137, 136, 137
Aston Martin motor cars: DB3, 121, 136; DB4, 121; DB5, 1, 2, 120, 165, 172, 185; gadgets in, 121–22, 125, 127; merchandise, 184–85
Atkins, Vera, 60
atomic bomb, 61, 73, 103, 111; as depicted in Bond films, 114–16; detection of explosions, 157–58, 162; fears of nuclear war, 108–9, 119; impact on Cold War, 110, 114; technological development, 110, 111–12
autogiro, 125, 168, 169

Baden-Powell, Robert, 16–17
Beatles, The, 122, 123
Bentley motor cars, 120, 121, 136, 140
Beretta firearms, 132–33, 134, 136
Big, Mr., 81, 139, 140
Blake, George, 147, 149, 154, 163
Bleriot, Louis, 12, 22, 29
Blofield, Ernst, 104, 105, 163, 184
Boer War, 8, 28
Bond, James, and ability to: drive automobiles, 9; maintain firearms, 14, 133; operate machinery, 111, 115, 186, 190
Bond, James, as a fantasy, 130, 164; as superhero, 128, 175, 176, 178; as surrogate for Ian Fleming, 118, 152, 163
Bond, James, as master of technology, 122, 141, 189; as modifier and improviser, 127, 133, 138, 140, 141
Bond, James, as modernizing hero, 2, 120, 122, 125, 189; saving the world 117; versus the machine, 3–4, 189, 190
Bond, James, duties of, 86, 131, 135, 137; marksmanship, 132, 133, 137; style, 186; unarmed combat, 92
Bond brand, 3, 75, 131, 184–88
Bondmania, 1, 51, 122, 129
Boeing company, 149, 170
Boothroyd, Major, 44, 132, 135
Braun, Wernher von, 40–41, 165
Broccoli, Albert ("Cubby"), 1, 61, 176, 177; and "bigger and better" Bond films, 116, 124–25, 126, 166; and Bond brand, 184; and gadgets, 123–24, 165, 172, 185; and "science fact," 171, 172, 185
Brosnan, Pierce, 1, 181–84
Bruce-Lockhart, Robert, 53, 54
Bruneval Raid, 61, 92–93
Buchan, John, 16, 18, 27, 54, 59
Buckmaster, Maurice, 90
Burgess, Guy, 69, 146, 147

card indexes, 25, 154
Carver, Elliot, 180, 182
Casino Royale (Fleming), 2, 9, 98, 108, 142, 176; and gadgets, 82, 139, 140
Casino Royale (2006 film), 2, 180
Chitty Chitty Bang Bang (Fleming), 175
Churchill, Winston, 5, 145, 164, 165; and air power, 11, 34, 38, 45, 46; career of, 6, 14, 18, 21, 128; and the Cold War, 142, 145; and irregular warfare, 28, 29, 64, 86; and nostalgia, 5, 6, 15, 163; and practical inventions, 66, 68–69, 73, 76; and science in the war effort, 34, 69; and secret service, 18–19, 26; and SOE, 64, 69; and technological enthusiasm, 7, 11, 14, 29, 34, 68; and threat of nuclear weapons, 108–9, 110

CIA, 44, 111, 129, 147, 148, 158; culture of, 161–62; formation of, 146; influence of Bond on, 128–29; and superiority of equipment, 155, 156; Technical Services Division, 128, 156, 157, 162; and technological advance in, 79, 125–26, 143, 160, 161–62
Clark, C. V. "Nobby," 77–78
Cloak and Dagger (film), 60–61
Clouds over Europe (film), 56, 57, 58
Cockleshell Heroes (film), 61–62
codebreaking, 21, 67, 73, 82; GC&CS, 65; GCHQ, 157
Cold War, the, 3, 108, 110, 117, 142, 145; as inspiration for Bond villains, 178, 180
Colt firearms, 13, 136; .45 "long barrel," 121, 136, 200; M1911, 136; Python, 200
computing, 119, 124, 155, 157, 177, 183; and personal computers, 181–82
concealment devices, 61, 71, 75–76, 80, 161, 176
Connery, Sean, 120, 125, 128, 170, 188; and gadgets, 134, 172, 173; on merchandising, 184
Cotton, Frederick Sidney, 33, 73; and Ian Fleming, 65–66; and innovation 33, 66; Sidcot flying suit, 33, 44
Cousteau, Jacques, 84–85, 126, 185, 187
Crabb, Lionel "Buster," 84, 130, 151
Craig, Daniel, 188
Cuban Missile Crisis, 108, 160
Cumming, Mansfield, 16, 19, 20, 26, 53, 64; and innovation, 24, 25, 34; and technological enthusiasm, 8, 11, 13, 24–25, 193

Dalton, Timothy, 181, 182
Dalzel-Job, Patrick, 88
Darko, Kerim Bey, 78, 96, 137
Davis Escape Apparatus, 83
death rays, 43–44, 48, 102, 110, 123
De Havilland aircraft: Be2, 30, 32; Comet, 149; DH2, 31
Deighton, Len, 151, 155, 170
Diamonds Are Forever (film), 93, 174, 178, 179, 180; and gadgets, 76, 111, 122, 170, 174; and lasers, 171; and space race, 170–71
Diamonds Are Forever (Fleming), 9, 93, 103, 132, 140, 152
Die Another Day (film), 171, 175, 176
Director of Naval Intelligence (DNI), 16, 21, 63
Doctor No (Fleming), 75, 81, 86, 95, 138, 148; and the Caribbean, 96, 166; and gadgets, 120, 133, 135, 138
Donovan, William "Wild Bill," 142–43, 144
Doyle, Arthur Conan, 29, 88
Drax, Hugo, 103–4, 105, 163, 180
Dreadnoughts, 14–15, 18, 20, 27, 28, 29
Dr. No (film), 1, 51, 75, 108, 111, 118–19; credits, 131; and firearms, 131–32, 133, 134; and gadgets, 75, 118, 127; set design, 119
Dr. Strangelove (film), 113
Drummond, Ace, 44
Drummond, Bulldog, 54, 56, 57, 86
Dulles, Allen, 128
Dunderdale, Wilfred "Biffy," 87–88

Eco, Umberto, 50, 176, 184, 189
Eden, Anthony, 77, 164
Edison, Thomas, 5, 34, 79
Edison films, 42, 50
Eichmann, Adolf, 75, 129
Elliot, Nicholas, 147, 149, 152
Eon Productions, 1, 121, 129, 166, 172, 177; and gadgets, 127, 129; and "science fact," 3, 172; and sets, 3, 119, 177
Everett, Douglas, 70, 71, 140
Evil, Dr., 116, 167

Fairbairn, William E., 91–92; Fairbairn and Sykes commando dagger, 78
Fantomas (film series), 51
FANY (First Aid Nursing Yeomanry), 87
Feuillade, Louis, 51
Flash Gordon (film series), 44, 45, 51, 103
Fleming, Ian, 5, 147, 165; and Bond's gadgets, 2, 10, 74, 76, 129–30; career of, 6, 15, 53, 192; character, 90, 94; and fears of terror weapons, 108, 112, 117; and Jamaica, 63, 94, 148, 166; and nostalgia, 1, 15, 67, 148–49; and patriotism, 163–64; and practical inventions, 65, 66–67, 73, 76, 139, 140–41; and relations with US, 142, 157
Fleming, Ian, and literary influences on, 50, 54, 80, 96; on writing Bond books, 54, 63, 85–86, 101, 142, 147
Fleming, Ian, and secret service, 53, 149, 152; as inspiration for Bond character, 92, 93–94, 96, 101–2, 106–7; in Naval Intelligence, 21, 63–65, 89–90, 95, 101; nostalgia about, 151–52, 163, 189
Fleming, Ian, and technological enthusiasm, 3, 29, 67, 111, 175; Fleming Collection, 9–10; on "terrible beauty" of V-2, 113, 184
Fleming, Peter, 64, 139
Foot, Michael, 87, 88, 139, 184
For Your Eyes Only (film), 95, 173; and gadgets, 126, 174, 175
For Your Eyes Only (Fleming), 67, 137
Frankenstein (films), 42–43

Fraser-Smith, Charles, 74–76
Frau im Mond (film), 40–41, 43, 167
From a View to a Kill (Fleming), 88, 137
From Russia with Love (film), 78, 95, 108, 119, 120; and gadgets, 1, 3, 120, 122, 137–38, 174
From Russia with Love (Fleming), 67, 82, 90, 95, 99, 152; and gadgets, 130; and Russian spies, 146–47; and "Spektor" machine, 97
From the Earth to the Moon (Verne), 39, 40
Frost, John, 93
Fuchs, Klaus, 99, 146
Fu Manchu, 50

gadgets, 2, 10, 74, 76, 123–24, 129–30, 165, 172, 185. *See also under names of specific films*
Geiger counters, 118–19, 124, 127, 156
Gestapo, 60, 75, 94, 103, 106
Get Smart (television series), 124
Goddard, Robert, 40, 151–52
Godfrey, John, DNI, 63, 64, 95, 142, 151
Goldeneye (film), 180, 181
Goldfinger, Auric, 104, 105, 106, 111, 185
Goldfinger (film), 2, 44, 93, 148, 180, 185; and atomic bomb, 111, 114; and gadgets, 121–22, 125, 127, 141, 156; and lasers, 44, 123, 171; and science, 171
Goldfinger (Fleming), 92, 93, 99, 121
Gordievsky, Oleg, 128
Gordon, Flash, 44, 49, 103
Grant, Red, 90, 120, 130, 137
Gubbins, Colin, 64, 72, 135

Handley-Page, Frederick, 8, 11
Handley-Page aircraft: 0/100, 30; 0/400, 30; V/1500, 36
Hannay, Richard, 18, 29, 54, 56, 67; influence on SIS, 87
Harmsworth, Alfred (Baron Northcliffe), 7, 11, 12, 18
harpoon guns, 138
Havilland, Geoffrey de, 13; and technological enthusiasm, 8, 11
Hawks, Graham, 126–27
helicopters: Bell, 168–69; Kawasaki, 169; Sikorsky, 168
Hildebrand Rarity (Fleming), 138
Hill, A. V., 36, 47, 69
Hitchcock, Alfred, 37, 54–56, 123, 195
Hitler, Adolf, 46, 48, 58, 80, 105, 200, as model for Bond villains, 104
Holdsworth, Gerry, 96–97
Holland, John Phillip, 14–15; submarines, 15, 29, 192
Holmes, Sherlock, 1, 17, 42, 54, 154
humor, 173–74, 181–82
hydrogen bomb, 110

I. G. Farben, 105–6, 116
inventions, practical, 65, 66–67, 73, 76, 139, 140–41
Ipcress File, The (Deighton), 155
irregular warfare, 28, 61–62, 65, 86–87

Jamaica, 63, 94, 148, 166
Jaws (film), 183
Jeffris, Millis, 69, 73, 77
Jones, Christmas, 55
Jones, R. V., 37, 47, 66, 98; and V-2, 101–2

Kennedy, John F., 128
KGB, 53, 157, 160, 199; Bond influence on, 128, 129
Klebb, Rosa, 3, 138
Kranzberg, Melvin, 109

Lang, Fritz, 40–41, 51–52, 60, 61, 165
Largo, Emile, 127, 140, 184
laser beams, 44, 123, 171–72, 174
Lawrence, T. E., 27, 28–29, 64
Le Carre, John, 147, 149, 151, 154–55
Le Chiffre, 88, 94, 98, 104, 139, 140
Leiter, Felix, 96, 148, 156, 161, 169, 182
Licence to Kill (film), 171, 173, 174, 180, 181, 182
limpet mines, 78, 80, 82–84
Lindemann, Frederick, 35, 46, 47
Live and Let Die (film), 173
Live and Let Die (Fleming), 81, 84, 95, 96, 108, 133
Living Daylights, The (film), 172, 181, 182
Llewelyn, Desmond, 1, 120, 173–74
Lockhart, Robert Bruce, 53, 54
Lotus Espirit, 125, 173, 185

M, 64, 103, 122, 151–52; on Bond's equipment, 131; briefing Bond, 96, 111, 148, 169
Mabuse, Dr., 51, 52
Mackenzie, Compton, 20, 53
Maclean, Donald, 146, 147, 153
Macrae, Stuart, 73, 74, 79, 80; development of limpet mine, 77–78
Man Who Knew Too Much, The (film), 54, 174
Man with the Golden Gun, The (film), 171, 174
March-Philips, Gus, 97
Marconi, Guglielmo, 8, 43
Mata Hari, 22, 51
Maugham, Somerset, 53, 54, 55

Maxim, Hiram, 13, 14
McNeile, H. C. ("Sapper"), 54, 55
MD1, research and development facility, 69, 74, 76, 78–79, 80; The Firs, 69, 74
Melies, Georges, 39, 165
Menzies, Stewart "C," 73, 88
merchandising, 184
Metropolis (film), 41–42, 119
MI5 (MO5), 26, 53, 63, 68, 153; formation of, 19; technological expertise, 153; as "tradesmen," 144
military technology, 13, 16, 134
Ming the Merciless, 44, 50–51
Minox miniature camera, 74–75, 160
Minshall, Merlin, 88
MI6, 26, 53, 63, 73, 112, 183; formation of, 19, 53; and relations with American intelligence, 146–47; and technology, 163. *See also* Secret Intelligence Service
Moonraker (film), 95, 108, 117, 172, 180; and gadgets, 75, 160, 171; and V-2, 184
Moonraker (Fleming), 102, 109, 141, 184
Moore, Roger, 169, 173, 181
Moore-Brabazon, John, 8, 11, 13, 149; and aerial reconnaissance, 22
motor torpedo boats (MTBs), 64, 82, 88
Muggeridge, Malcolm, 144–45, 147, 155
Mumford, Lewis, 5, 10, 48, 109

NASA, 40, 41, 106, 166, 170, 179; Apollo program, 170; Gemini program, 166; Mercury program, 169; Space Shuttle, 117, 172
Naval Intelligence Department (NID), 16, 63, 74
nerve agents, 95, 99, 105–6, 116–17
Never Say Never Again (film), 134
Nikon cameras, 185, 186
No, Julius, 104, 106, 111, 139, 141
North by Northwest (film), 169
nostalgia, 1, 15, 67, 148–49; about secret service, 151–52, 163, 189
nuclear weapons, 108–9, 110

Oberth, Herman, 39–41
Octopussy (film), 95, 108, 114, 173; and gadgets, 115, 133, 134, 169, 175
Octopussy (Fleming), 87, 89, 98
On Her Majesty's Secret Service, (film), 1, 75, 116–17, 170
OSS (Office of Strategic Services), 59, 138, 143; and "Oh So Social," 144, 161
Our Man Flint (film), 123–24
Our Man in Havana (film), 162

parachutists, 61, 88–89
patriotism, 163–64
Pearson, John, 63, 64, 74
Penkovsky, Oleg, 159–60, 161, 162, 202
Philby, Kim, 87, 123, 147–48, 150, 201; and Ian Fleming, 129, 147–48; and relationship with American intelligence, 145, 146; on Secret Intelligence Service, 129, 148, 153, 154; as "Third Man," 146–47
photographic interpretation, 23, 66–67, 160
photography in espionage, 20, 25, 52, 74–75, 161
Playboy, 124, 186
poison gas, 27, 44, 45, 102, 196
poisons, 138–39, 143, 196
Polaris missiles, 114, 190
Porton Down chemical warfare center, 36, 138
product placement, 34, 45, 67, 184

Q, 90, 129, 134, 147, 157, 171; as humorous character, 173–74, 181–82; origins of, 74, 76; popularity with film audiences, 120; relations with Bond, 1
Q Branch, 118, 120, 138
Quantum of Solace (film), 135
Quarrel, 96, 118

radar, 47, 48, 61, 73, 83
Red Beret (film), 61
Reilly, Sidney, 12, 17, 30, 67; "Ace of Spies," 53–54
Remington firearms, 139
Rickenbacker, Eddie, 31, 44
Rider, Honeychile (Ryder, Honey), 1, 104
Rogers, Buck, 44–45, 49
Rolls, Charles, 8, 11, 13
Rolls Royce motor cars, 9, 12, 27
Royal Air Force (RAF), 30, 34, 46, 64, 104, 168; and weapons research, 36
Royal Flying Corps (RFC), 22–24, 30
Royal Naval Air Service (RNAS), 11, 23, 33, 34
Royal Navy, 6, 14–15, 20, 114
Rutherford, Ernest, 34–35

Saltzman, Harry, 1, 176, 177; and "bigger and better" Bond films, 116, 124–25, 126, 166; and Bond brand, 184; and gadgets, 123–24, 165, 172, 185; and "science fact," 171, 172, 185
Savage firearms, 137
science in the war effort, 33, 34–35, 47–48, 68, 70–71; and the Nazis, 101, 106–7; and "The War of Experts," 73
Second Industrial Revolution, 2, 5, 7, 12; and espionage technology, 16–17, 20; and military technology, 13, 16, 134

Secret Agent (film), 37, 55–56, 58
secret inks, 20, 24–25, 34
Secret Intelligence Service (SIS), 24–25, 53, 144, 147; aviation section of, 46; culture of, 148, 149, 150, 151, 154; influence of Bond, 138; relationship with American intelligence, 144–47; Section D of, 69; and technical advance, 26, 149–50, 153, 163. *See also* MI6
Shape of Things to Come, The (Wells), 15
Shatterhand, Guntram, 105, 145
SIGINT (signals intelligence), 21, 26, 27, 146, 156–57, 162–63
Skyfall (film), 135, 180, 188
Sky Hook (rescue system), 125–26
SMERSH, 44, 137, 176
Smith & Wesson firearms, 132, 133
Smythe, Major, 87, 98, 99, 132, 139
SOE research and development stations, 69–70, 135–36; Station IX, "The Frythe," 70, 78, 80, 83, 140; Station XII, Aston House, 70; Station XIV, 70; Station XV, "The Thatched Barn," 71–72; Station XVII, Hertford, 72
Sony Corporation, 170, 178, 186
sound recording in espionage, 20, 25, 122–23, 153; in Bond films, 122
Special Operations Executive (SOE), 2, 59, 60, 170, 184; formation of, 64; training, 89–91
SPECTRE, 90, 140
Spectre (film), 189
Spione (film), 52, 167
Spy Who Loved Me, The (film), 95, 108, 172, 173, 180, 183; and gadgets, 125, 175, 185, 187, 190; and poison gas, 117; and underwater technology, 114; villains in, 177, 180
Star Wars (film), 60, 183
Sten gun, 118, 135–36
Stephenson, William, 8, 9, 30; and aerial reconnaissance, 21; and BSC, 142–43; and Ian Fleming, 8, 9
strategic bombing, 45–46, 111–12; in World War I, 30, 32, 38; in World War II, 99
Stromberg, Karl, 108, 177, 178, 180
submarine detection technology, 34–35, 114, 173
submarines, 14–15, 16, 18, 64; human torpedoes, 29, 83–85; midget, 29, 83, 100; nuclear-powered, 113–14; in World War I, 27, 29. *See also* Holland, John Phillip; U-Boats
surveillance technology, 53, 159–60, 163
Sykes, Eric A., 91, 92

Tanaka, Tiger, 145, 169
terror weapons, 41, 46, 52, 99–102, 109. *See also* Zeppelins
Things to Come (film), 45, 46, 49
30 Assault Unit, 84, 95, 100–102; formation of, 95; as piracy, 98; training, 97–98
Thirty-Nine Steps, The (Buchan), 18, 55
39 Steps, The (film), 55–56
Thunderball (film), 1, 58, 95, 111, 112, 173; and atomic bomb, 114, 184; and gadgets, 78, 94, 122, 125, 156, 165; and Russian spies, 146; and underwater technology, 126, 127–28, 138, 185
Thunderball (Fleming), 82, 95, 103, 105, 112, 152; and *Olterra*, 84–85; and underwater technology, 127
time fuses, 77–78, 80
Tizard, Henry, 35, 47–48
Tomorrow Never Dies (film), 134, 180–81, 182
torture, 61, 94, 104
Toyota automobiles, 167, 170, 185
Trip to the Moon, A (film), 39
20,000 Leagues under the Sea (films), 39, 119
20,000 Leagues under the Sea (Verne), 39, 113, 177

U-Boats, 20, 21, 63, 88; XXI, 113; Walter, 100
Ultra Secret, 67, 73
unarmed combat, 91–92
underwater warfare, 29, 84–85

V-1 cruise missile, 100, 101, 102
V-2 ballistic missile, 41, 100, 101–2, 109, 158, 184
Verne, Jules, 29, 39, 165
View to a Kill, A (film), 95, 177, 178, 180, 181, 183
View to a Kill, A (Fleming), 151
villains, 104, 177, 178, 180. *See also names of specific people*

Walter, Helmut, 100
Walther firearms, 132–33; Model 1, 134; P5, 134; P99, 134–35; PPK, 133–34, 140
War in the Air, The (Wells), 7, 15, 16, 33, 41, 45
War of the Worlds (Wells), 40, 43
watches, 186–87; Breitling, 127; Rolex 173, 187; Seiko, 115, 175, 187
Watson-Watt, Robert, 35, 47–48
Webley firearms, 132
Wells, H. G., 12–13, 14, 33, 39, 40, 45
Wimperis, H. E., 47, 48
Winder, Simon, 62, 164, 165, 183
Winterbotham, Frederick, 7; and aerial reconnaissance, 23, 65–66; and military aviation, 23, 30, 32; and secret intelligence, 46–47; and technological enthusiasm, 7–8, 10, 14; and Ultra Secret, 73

wireless communication in espionage, 21, 80–81, 161, 170
World Is Not Enough, The (film), 108, 115, 117, 174, 180, 183; and gadgets, 137
Wright, Orville and Wilbur, 6, 7, 10–11, 12, 39
Wright, Peter, 149–50, 153, 163; and The Thing, 159
Wynne, Greville, 129, 160, 162, 202

X-craft (submersibles), 93

Yankee Doodle in Berlin (film), 51–52
You Only Live Twice (film), 117, 166, 170, 172, 180; and gadgets, 165, 168, 169, 185, 186; and sets, 167
You Only Live Twice (Fleming), 105, 147

Zeppelin, Ferdinand von, 6, 12
Zeppelins, 12, 15, 16, 18, 27, 34; attacks on England, 32, 37–38
Zorin, Max, 180